Hemodynamic Aspects of Cerebral Angiomas

Werner Hassler

Acta Neurochirurgica
Supplementum 37

Springer-Verlag Wien New York

Werner Hassler, M.D.

Department of Neurosurgery, University of Tübingen, Federal Republic of Germany

Translated by Helmuth Steinmetz, Tübingen, Federal Republic of Germany

With 137 Figures

Library of Congress Cataloging-in-Publication Data. Hassler, W. (Werner) Hemodynamic aspects of cerebral angiomas. (Acta neurochirurgica. Supplementum; 37) Bibliography: p. 1. Brain—Blood-vessels—Surgery. 2. Hemangioma—Surgery. 3. Hemodynamics. 4. Brain—Blood-vessels—Radiography. 5. Ultrasonic encephalography. 6. Surgery, Experimental. I. Title. II. Series. [DNLM: 1. Brain Neoplasms—blood supply. 2. Hemangioma—blood supply. W1 AC8661 no. 37 / WL 358 H355h] RD594.2.H39 1986. 616.99′281. 86-30028.

ISSN 0065-1419

ISBN-13:978-3-211-81970-8   e-ISBN-13:978-3-7091-8891-0
DOI: 10.1007/978-3-7091-8891-0

To Mechthild,
Sabine and Christine

Foreword

Up to date, the treatment of arteriovenous racemose angiomas of the brain remains unsatisfactory. Intraoperative hemorrhages, post-embolizational or postoperative deficits depending on the site and size of the AVM as well as inoperability of rare angioma types have promoted the technical improvement of diagnostic and therapeutic approaches. Nevertheless, some pathophysiological problems of AVM hemodynamics have not been solved. Many angiographical studies, observations during embolization and operation, dopplersonographical and other perfusion measurements provided some insight. Sufficient animal models have yet to be developed in order to elucidate the pathophysiological mechanisms. This monograph describes AV fistula models in cats and rats, both conventional and newly developed, which allow a better comparison with human cerebral angiomas than previous ones. The most important result is that the model of the breakthrough of arterial pressure waves into the capillaries following a failure of cerebrovascular regulation cannot be confirmed. Rather, according to the findings in precapillary vessels presented here, the regulation functions normally so that a breakdown of regulation cannot be responsible for global brain edema often seen after removal of angiomas. The regulation was demonstrated using different methods, most important of which being the CO_2 response of brain vessels to varying CO_2 contents of the inhaled air. Angiographical, dopplersonographical and perioperative dopplersonographical as well as intraoperative measurements of flow and pressure have been applied. Angiographic and transcranial Doppler findings concur following removal of angiomas and fistulas in humans and animal models. The conclusion was reached that pressure passive rupture of thin-walled vessels must be seen as the cause for intra- and postoperative bleeding. As it was the case with other authors, no sufficient method was available to detect the pressure drop between arterioles and the venous part of the capillary system. The question remains to what extent minimal increases in this pressure gradient play a role in the development of edema since regulation was proved solely by the CO_2 response, and it is unknown whether minimal restrictions of regulation exist which may nevertheless be hemodynamically relevant. Finally, the characteristics of large angioma feeders are studied. Their calibers initially remain enlarged in spite of preserved regulation. At the same time, brain vessels, especially the larger segments like the carotid, react within a few hours to the exclusion of arteriovenous fistulas showing a sensible constriction of vessel walls in humans and animals as well. The question of arterial or venous thromboses is also discussed.

For the clinician, the results mean that he has to focus his operative technique even more than before to the preservation of major inflows and outflows of the angioma until the end of the operation and occlude the many small feeders at a moment when pressure-dependent rupture can be avoided because large shunts are still patent. The main problem with postoperative hemorrhage is not the rupture of larger vessels but that of small ones which should be adequately thrombosed at the end of the operation. Application of this modern technique should reduce risks of postoperative bleeding. The microsurgical experience as accumulated in the past few years by several modern centers strongly suggests this approach. However, some questions remain open such as whether large angiomas should be operated upon in several stages in short intervals as recommended by Spetzler. On the contrary, large malformations can be removed in one session in spite of increased risk of edema. This has been Yaşargil's practice up until now since he quite rightly anticipates the possibility of hemorrhage from residual angioma parts as well as the rapid enlargement of shunts that have been left in place and that may reach considerable size within a few weeks. Further problems persist which cannot be solved in this context.

The main value of this study is the fact that it has thrown light on the pathophysiology of the condition and thereby facilitates further therapeutic progress.

Freiburg i. Br., December 1986   Wolfgang Seeger

Foreword

Repeated transcranial Doppler sonography of intracranial vessels has been proven reliable and useful in investigating pathophysiology after subarachnoid bleeding and trauma. Both in research and in clinical work our basic understanding of the reactions of the cerebral vessels has been increased and has eased clinical decision making.

W. Hassler has introduced this method for studying blood-flow phenomena in cerebral and spinal angiomas and fistulas. Animal experiments have shown the method applicable in investigating shunts of the central nervous system. The findings in man demonstrate with great reliability and specificity typical rheological situations of the angiomas, differentiating between those feeders only supplying the brain, those supplying exclusively the angioma and those with mixed function.

Doppler sonography may work as a safe, unrisky pre-, intra- and postoperative monitoring system in angioma-surgery, whereas repeated angiography would add a big load to the patient's situation.

After introducing Hassler's findings and experience into clinical work we learned that cerebral vasculature is reacting within minutes to different conditions, e.g. CO_2 level, blood pressure, intracranial pressure, and that angiographic findings are reflecting only a moment's situation. The necessity of frequent investigation in critical cerebral vascular situations is very obvious and now possible without any harm to the patient.

Most important in the intra- and postoperative course after resection of a cerebral AV-angioma are the facts that CO_2 reactivity of the feeding vessels is preserved and that perfusion of the affected hemisphere drops dramatically, much below normal circulation and that of the contralateral hemisphere. We now know that hypervolemia does not exist in the postoperative phase and that complications in form of hematomas are not due to breakthrough but to incomplete resection or rupture of small feeders.

Tübingen, December 1986   Ernst H. Grote

Acknowledgements

I would like to take this opportunity to express my gratitude to my teacher Prof. Dr. Wolfgang Seeger who fostered my interest in the hemodynamics of cerebral angiomas and generously afforded me time for the studies. Special thanks also to my present chief, Prof. Dr. Ernst H. Grote who made possible the continuation of my scientific work, to my collaborator Dr. Helmuth Steinmetz and to Dr. Alec Eden for finishing touches as well as to Virginia Müller for typing the manuscripts.

Furthermore, I would like to thank the following: my doctoral candidates Mr. Eble and Mr. Koppermann for their contributions to the experimental part, Prof. Meureth (Anaesthesiology) for entrusting me with the electromagnetical flowmeters, Dr. Blumberg (Neurophysiology) for providing the laser Doppler, Prof. Spatz (Brain Research) and Dr. Zöllner (ENT) for making the operating room available. Professor Volk (Neuropathology) and Prof. Staubesand (Anatomy) are thanked for the electrone microscopic evaluations, Dr. Birg for his mathematical assistance, Mr. Ebner for the construction of special instruments, Dr. Roth and Mr. Schuler for taking care of the animals as well as Mr. Pfister, Mrs. Huber and Mr. Eiben for their wonderful photographic work.

Tübingen, December 1986   Werner Hassler

Contents

I. **Introduction**. 1
    1. History 1
    2. Aim of the Study 2
    3. Hemodynamic Principles of Cerebral Perfusion 2
        3.1. Hemodynamics in Normal Blood Vessels 2
        3.2. Cerebral Perfusion and Autoregulation 3
    4. Hemodynamics in Cerebral Angiomas 5
        4.1. Hemodynamics of Angioma Feeding Arteries 5
        4.2. Hemodynamics in the Angioma 5
        4.3. Brain Perfusion and Angioma 5
        4.4. Hemodynamic Changes After Angioma Exclusion 6

II. **Animal Experiments** 7
    1. Experimental Arteriovenous Fistulae 7
        1.1. The Fistula Models of Spetzler and of Scott 7
        1.2. Own Fistula Models 7
        1.3. Topics of Our Experimental Investigations 8
    2. Materials and Methods 9
        2.1. Anaesthesia 9
        2.2. Surgical Procedure 9
        2.3. Blood Pressure Measurements 9
        2.4. Electromagnetic Measurements of Flow Rate 9
        2.5. Dopplersonographic Measurements of Flow Velocity 10
        2.6. Measurements of Microcirculation (Laser Doppler) 10
        2.7. Angiography 10
    3. Results of the Fistula Models in Rats 10
        3.1. Direct AV Fistula (T-fistula) 10
            3.1.1. Electromagnetic Flow Measurements 10
            3.1.2. Intraoperative Doppler Sonography 12
            3.1.3. Blood Pressure 12
            3.1.4. Angiography 12
        3.2. Indirect AV Fistula (H-fistula) 13
            3.2.1. Electromagnetic Measurements 14
            3.2.2. Doppler Sonography 15
            3.2.3. Blood Pressure Measurements 15
            3.2.4. Angiography 16
            3.2.5. Long-term Experiments 16
            3.2.6. Autoregulation Experiments 18
    4. Results of Fistula Models in Cats 18
        4.1. Flow Rates and Blood Pressure in the H-fistula Model . . . 19
        4.2. Dopplersonographic Measurements 20
        4.3. Measurements in Varying Fistula Flow 20

    4.4. High Flow Velocities and Vessel Wall Vibrations 22
    4.5. Autoregulation in Acute and Chronic H-fistulas 22
    4.6. Measurements of Cortical Perfusion and Flow Rates 24
    4.7. Blood-brain Barrier 28
  5. Discussion and Summary of Animal Experiments 29

III. **Angiography in Angioma Patients Before and After Surgery** 33
  1. Materials and Methods 33
  2. Results 34
  3. Discussion 35
  4. Summary and Conclusion 36

IV. **Transcranial Doppler Sonography in Angioma Patients** 38
  1. Introduction, Methods and Normal Values 38
    1.1. Transcutaneous Recording Methods 38
    1.2. Transcranial Doppler Sonography 39
    1.3. Normal Values 42
    1.4. Discussion (Normal Values) 43
  2. Transcranial Doppler Sonography in Angioma Patients (Normocapnia) 45
    2.1. Preoperative Recordings 45
      2.1.1. Flow Characteristics of Angioma Supplying Arteries 45
      2.1.2. Discrimination of Spastic and Angioma Feeding Arteries 49
      2.1.3. Transcranial dopplersonographic Recordings and Angiographical Vessel Diameter . 49
      2.1.4. Altered Blood Distribution and Steal Effects 57
      2.1.5. Compression Tests in Angiomas 57
      2.1.6. Special Cases 58
      2.1.7. Flow Characteristics of Angioma Veins 60
      2.1.8. Flow Characteristics of Cerebral Arteries 60
    2.2. Postoperative Recordings in Angioma Patients 60
      2.2.1. Flow Characteristics in Former Angioma Feeders 60
      2.2.2. Flow Characteristics in Neighboring Brain Arteries After AVM Removal . . . 74
      2.2.3. Special Case With Early Hemorrhage After Surgery 74
    2.3. Discussion 77
    2.4. Summary 79
  3. CO_2 Reactivity, Normal Values and Findings in Angioma Patients 79
    3.1. Method 80
    3.2. Terminology 82
    3.3. Normal Values 82
    3.4. CO_2 Reactivity in Angioma Patients 87
    3.5. CO_2 Reactivity of Brain Arteries in the Angiogram 93
    3.6. CO_2 Reactivity After AVM Removal 94
    3.7. CO_2 Reactivity Before and After Partial Embolization 96
    3.8. Special Cases 96
    3.9. Discussion 101

V. **Intraoperative Studies in Angioma Patients** 109
  1. Intraoperative Methods (History) 109
  2. Topics of Our Intraoperative Measurements 110
  3. Methods Used in This Study 110
    3.1. Microvascular Intraoperative Doppler Sonography 110
    3.2. Electromagnetic Measurements of Flow Rate 110
    3.3. Intravasal Pressure Measurements 110

4. Operative Strategy and Preparation of Vessels to Be Measured 111
5. Measurements Before AVM Removal 111
   5.1. Doppler Sonography Before AVM Removal 112
   5.2. Electromagnetic Measurements of Flow Rate 112
   5.3. Intravasal Pressure Measurements 112
6. Measurements After AVM Removal 113
   6.1. Doppler Sonography After AVM Removal 113
   6.2. Intravasal Pressure Measurements After AVM Removal 115
   6.3. Electromagnetic Measurements 117
7. CO_2 Reactivity of Vessels Supplying Angioma and Brain Before and After AVM Exclusion . 117
   7.1. CO_2 Reactivity Before AVM Removal 119
   7.2. CO_2 Reactivity After AVM Removal 119
8. Discussion 122
9. Summary of Intraoperative Measurements 125

VI. **Final Conclusions** 126

References 127

Subject Index 135

List of Abbreviations

A 1	horizontal part (A 1 segment) of ACA		P 2	P 2 segment of PCA
ACA	anterior cerebral artery		pCO_2	partial pressure of CO_2
AVM	arteriovenous malformation		PCA	posterior cerebral artery
BP	blood pressure		PICA	posterior inferior cerebellar artery
CBF	cerebral blood flow		RI	resistance index
CCA	common carotid artery		SABP	systemic arterial blood pressure
ICA	internal carotid artery		VJC	common jugular vein
MCA	middle cerebral artery		VJE	external jugular vein
P 1	P 1 segment of PCA		VJI	internal jugular vein

All transcranial dopplersonographical measurements were done
with a 2 MHz pulsed Doppler divice, so that
1 kHz of Doppler shift corresponds to 39 cm/sec of flow velocity.

I. Introduction

1. History

A cerebral angioma was described for the first time in 1757 by Hunter. The first pathoanatomical studies were done in 1863 by Virchow, who at the time thought he was dealing with angiomatous tumors. In 1888 Powers recognized them to be congenital vascular malformations not originating from the brain tissue itself.

Cushing and Bailey were the first to explore operatively an angioma while searching for an epileptic focus in 1921. Removal did not enter their minds, as they lacked the knowledge of the pathoanatomical nature of the process. Together with Dandy they later described angiomas in man as vascular malformations which are surrounded by thin arteries and which have red veins running from it. They were the first to recognize arteriovenous shunt as a hemodynamic characteristic of angiomas.

After its introduction by Moniz, cerebral angiography was utilized by Olivecrona to localize preoperatively angiomas with feeding arteries and draining veins. Angiography demonstrated fast blood flow in the malformation itself, early filling of veins and relatively scarce filling of the surrounding brain vessels.

The era of angioma surgery started in 1932 when Olivecrona successfully carried out his first radical removal of a cerebral angioma. In 1948 he published his experience with surgical treatment. Three out of 24 patients had died in his series. One year later, Norlen reported 10 complete angioma removals without any mortality (Norlen 1949). His fundamental observation was that two or three weeks after surgery the enlarged vessels returned to normal size whereas previously poor filling of the surrounding brain vessels improved.

In 1948 Shenkin (using the Kety-Schmidt method) demonstrated threefold flow velocity from the internal carotid artery to the internal jugular vein as well as a 30–122% increase of cardiac output in angioma patients. This was the beginning of clinical research on hemodynamics in angiomas.

Six years later Murphy found intracerebral steal phenomena leading to impaired perfusion of normal brain regions. This was connected with clinical findings such as epileptic seizures and psychic alterations (Murphy 1954).

Further knowledge on hemodynamic changes was facilitated by measurements of regional cerebral blood flow (Ingvar and Lassen 1962, Feindel 1967, Harper 1965, Symon 1968). However, new hemodynamic concepts did not result from such sophisticated methods.

The first intraoperative measurements were done by Feindel in 1971. Using xenon 133 and fluorescin angiography he found that steal phenomena decreased and perfusion of the surrounding brain improved after exclusion of angiomas.

The most important findings were those of Nornes (1979, 1980). He measured blood flow velocities intraoperatively with a pulsed Doppler device and also registered perfusion pressure. Pressure was found to rise up to 50% after angioma exclusion.

The surgical techniques were also refined. Many of the angiomas formerly considered inoperable became accessible due to progress in microsurgical technique and instrumentation (Wilson 1979, Drake 1979, Yaşargil 1976, Seeger 1980, 1985). However, time and again major complications occurred and were explained by hemodynamic changes following angioma exclusion.

Most of the problems consisted of swelling and bleeding (Drake 1979, Mullan 1979, Wilson 1979), which were attributed to postoperatively elevated perfusion pressure. In order to prevent this, many therapeutic solutions have been attempted. Some authors reduced the size of angiomas by preoperative partial embolization (Cophignon 1978, Deruty 1981, Debrun 1982, Luessenhop 1982), others postoperatively lowered systemic blood pressure (Pertuiset 1981, Luessenhop 1982), recommended step-wise removal (Wilson 1979) or tried to lower local perfusion pressure by throttling the carotid with Selverstone clamps

(Pertuiset 1983 and Bonnal 1985). Another concept was the electro-physiological inactivation by barbiturates (Day 1982).

The assumed impact of postoperatively raised perfusion pressure is labelled by such terms as "circulatory breakthrough" (Nornes 1977), "normal perfusion pressure breakthrough theory" (Spetzler 1978) and "proximal hyperemia" (Mullan 1979). The common conception of these authors is that of impaired vascular autoregulation in the vicinity of an angioma. Sufficient postoperative vasoconstriction of adjacent brain arterioles is thought to be impossible after long-standing dilation prior to the operation. Accordingly, brain swelling and bleeding would result (Spetzler 1978).

Until now, angiography was the only way to recognize preoperatively hemodynamically critical angiomas. These were considered to show poor filling of surrounding brain arteries, steal phenomena, high flow shunts and feeders of more than 8 cm length (Walter 1975, Wilson 1979, Drake 1979, Luessenhop 1982 and Nornes 1975, Hassler 1984). To date, other criteria for hemodynamical assessment of angiomas are lacking.

2. Aim of the Study

The aim of this study was the evaluation of new methods for hemodynamic assessment of arteriovenous malformations (AVM). In addition, the present concepts of hemodynamic changes after AVM exclusion were to be reconsidered.

In the *experimental part* the attempt was made to find out

1. simple methods to detect hemodynamic characteristics of AV fistulas,

2. vascular hemodynamic models for angioma vessels,

3. vascular models for brain supplying vessels leaving the angioma feeders,

4. the effect of acutely and chronically reduced perfusion pressure on the brain as well as the vascular autoregulative capacity under these conditions.

In the *clinical part* of this investigation the attempt was undertaken to evaluate

1. methods to detect hemodynamic characteristics of arteries supplying the brain and angioma; with these, the hemodynamic impact of angiomas and implications for assessment of adequate therapy;

2. hemodynamic changes following angioma exclusion, their significance for postoperative complications

and the time needed for adjustment to normal conditions;

3. the reactivity of resistance vessels supplying brain and angioma to pCO_2 changes before and after angioma exclusion;

4. intraoperatively measured flow and pressure of AVM-feeders as well as vasomotor response to pCO_2 changes before and after angioma exclusion.

3. Hemodynamic Principles of Cerebral Perfusion

3.1. Hemodynamics in Normal Blood Vessels

The rate of blood flow Q depends on the difference in pressure between both ends of a given vessel (Δp) and on the resistance R to that flow. Only 2% of heart action is transformed into kinetic energy; 98% serves to overcome the frictional force (Nornes 1980). As in other parenchymatous organs, this resistance in brain circulation is low compared to relatively high resistance in muscles (extracranial circulation).

The following formula applies:

$$Q = \frac{\Delta p}{R} \qquad (1)$$

Resistance R to blood flow is given by

$$R = \frac{\Delta p}{Q} \qquad (2)$$

R increases in proportion to blood viscosity η which results from frictional forces between the concentric fluid layers.

$$\eta = \frac{\tau}{\gamma} \qquad (3)$$

τ = shearing force and γ = gradient of velocity between the layers

The viscosity of blood is inversely related to temperature. It is indicated in relative units with the basic value being 1.0 for water. Only homogenous fluids show constant viscosity whereas inhomogenous fluids such as blood have variable viscosity changing with velocity and red cell concentration. Very low flow velocities can result in relative values of more than 1,000 ("prestasis"), which, under physiological conditions, can only occur in very small vessels.

The Fahraeus-Lindquist effect occurs in vessels of less than 1 mm diameter. In these vessels the erythrocytes join in the central axis, whereas in the cell-free marginal zone viscosity is lowered to 50% of that found in larger vessels (Schmidt, Thews 1985).

With the exception of very large vessels such as the aorta, blood flow shows a parabolic profile of velocity with its maximum in the center and peripherally decreasing velocities. In this so-called laminar flow pattern, all particles move parallel to the vessel's axis.

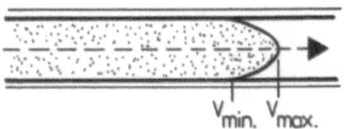

Fig. 1. Parabolic velocity profile in laminar flow

The flow velocity under laminar conditions is

$$v = \frac{\Delta p \cdot r^2}{8 \cdot l \cdot \eta} \qquad (4)$$

v = mean velocity, Δp = pressure gradient, r = vessel's radius, η = viscosity and l = vessel's length

The rate of blood flow is

$$Q = v \cdot \pi \cdot r^2 \qquad (5)$$

The combination of formulae (4) and (5) results in Hagen-Poiseuille's Law

$$Q = \frac{\Delta p \cdot \pi \cdot r^4}{8 \cdot l \cdot \eta} \qquad (6)$$

Therefore applies:

$$R = \frac{8 \cdot l \cdot \eta}{\pi \cdot r^4} \qquad (7)$$

Hagen-Poiseuille's Law implies that flow rate and resistance change in direct or inverse proportion to the fourth power of the diameter. Therefore, the vessel's diameter is the crucial factor in circulatory regulation.

The applicability of these physical laws to biological systems is limited by the following factors. Hagen-Poiseuille's Law is only effective in laminar flow patterns, rigid tubes of constant diameter and homogenous fluids. Blood is an inhomogenous fluid with pulsatile flow in elastic vessels of changing diameter. In addition, pathological conditions such as AV fistulas may lead to turbulent flow. This must be considered in theoretical models of hemodynamics. The rate of turbulent flow, for example, is no longer linearly related to a vessel's pressure gradient Δp, but to its second power. To double flow rate under this condition, fourfold pressure is necessary.

Blood flow is not only influenced by resistance but also by the elasticity of vessels. A well-known example is the "windkessel" effect of the aorta. Elasticity is determined by the proportion of elastic and collagenous fibers. Elastic properties of a hollow organ are characterized by the volume elasticity coefficient E'

$$E' = \frac{\Delta P}{\Delta V} \qquad (8)$$

ΔP = change of pressure, ΔV = change of volume

In this formula, low E' corresponds to high elasticity.

High flow velocities as well as vessel stenoses result in turbulencies. They occur when the so-called Reynolds' Value (Re) is exceeded. Values over 400 appear with marginal whirls. Above 2.000, laminar flow is completely abolished.

$$Re = \frac{r \cdot \bar{v} \cdot \rho}{\eta} \qquad (9)$$

r = radius, $\bar{v}$ = mean velocity, ρ = volumetric weight (kg/m), η = viscosity

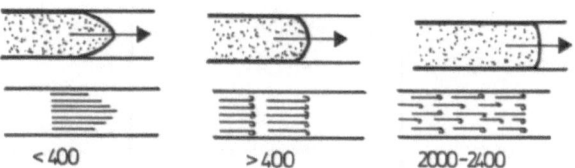

Fig. 2. Flow profiles with varying Reynolds' values

Values of 400 are reached when flow velocity increases to 102 cm/sec. 38.1 ml/min then would have to pass through a vessel of 1 mm diameter. A Reynolds' Value of 2.000 would require a flow of 239.9 ml/min through the same vessel and a velocity of 509 cm/sec, which have never been found in brain vessels even under pathological conditions.

3.2. Cerebral Perfusion and Autoregulation

Cerebral blood flow (CBF) amounts to 750 ml/min, which is 13% of the cardiac output volume. CBF of grey matter is 0.8–1.1 ml/g/min, that of white matter 0.15–0.25 ml/g/min. This results in a mean CBF of 0.5 ml/g/min (Schmidt, Thews 1985).

Total CBF remains stable even with very different functional states such as consciousness or sleep. It drops, however, in general anesthesia, hypothermia and metabolic coma. A rise in total CBF is seen in epileptic seizures and hyperthermia. Depending on different mental activities, regional CBF can vary, leaving total CBF unchanged (Siesjö 1980).

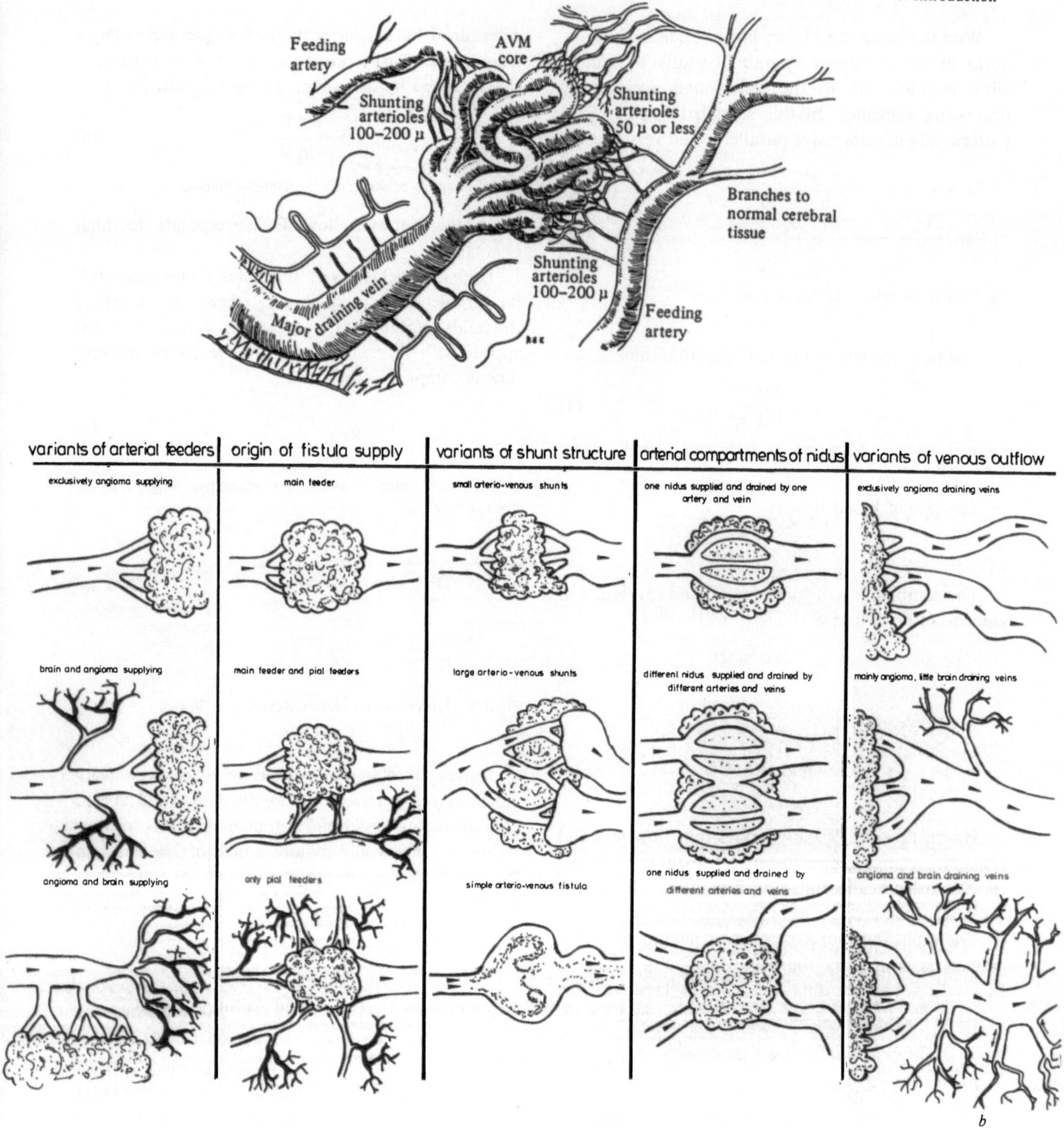

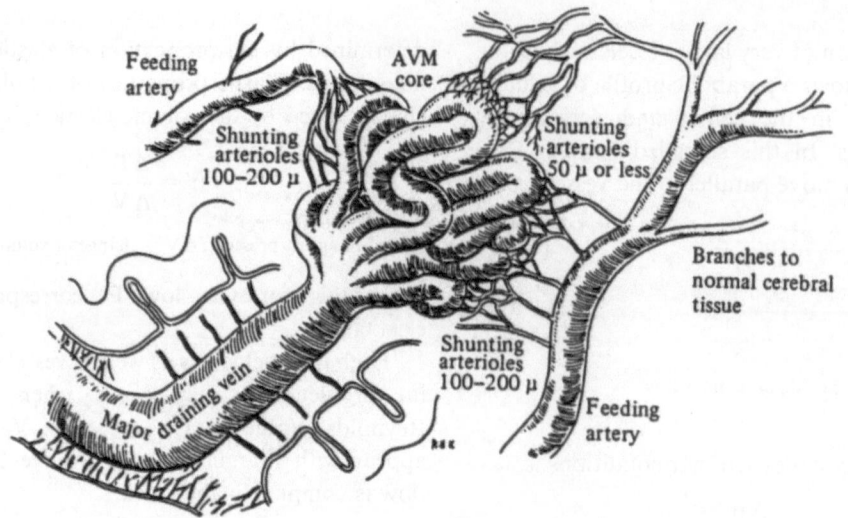

Fig. 3 a and b. a) Schematic drawing of an angioma (Yamada 1982) b) Angioarchitecture of AV-malformation as seen in angiography and during embolization (modified after Rüfenacht, D.; J. J. Merland; A. Laurents; 1986): Varying types of AVM-feeders, origin of fistula supply, varying shunt structures, different compartments of AVM nidus, and varying types of venous outflow

Cerebral vascular autoregulation is based upon myogenic changes in the diameter of the arterioles (resistance vessels). Thus, total CBF remains stable throughout wide ranges of blood pressure. The lower limit of this range is reported to be about 50 mmHg of mean arterial pressure for normotensives. The upper limit is at 150 mmHg. In the experimental animal, no upper limit was found under normocapnic conditions. Hypercapnia caused a drop to 125–150 mmHg, depending on the level to which pCO_2 was raised. This shows that the autoregulative capacity can be reduced by pCO_2 changes (Lassen 1972).

Cerebral perfusion is mainly controlled by metabolic factors. Most important are the levels of CO_2 in blood and tissue as well as perivascular H^+ concentration. A rise in pCO_2 leads to noticeable arteriolar vasodilatation followed by an increase in CBF. The same happens in brain tissue acidosis. Falling pCO_2 induces vasoconstriction and hypoperfusion. Changes in pO_2 do not influence CBF.

4. Hemodynamics in Cerebral Angiomas

The angioma itself consists of primitive, altered arterioles which directly shunt into venous loops. These shunting arterioles measure 50–200 μ in diameter (Parkinson 1958) and branch off the so-called AVM feeding arteries. The venous loops (0.5–2 mm) run into the main draining veins. It is important to mention that the adjacent brain tissue is supplied by normal arterioles also issuing from the AVM-feeding artery. Small arterioles and venules connect angioma loops with normal brain vessels (Aronson 1972 and McCormick 1966).

Our idea of hemodynamics in cerebral angiomas and its effect on the surrounding brain tissue is vague. Two authors (Spetzler 1978, Nornes 1980) have worked on these phenomena before and after removal of angiomas. Their concepts are summarized as follows:

4.1. Hemodynamics of Angioma Feeding Arteries

Nornes clearly elucidated some basic physical principles in his experimental model of angioma feeding arteries.

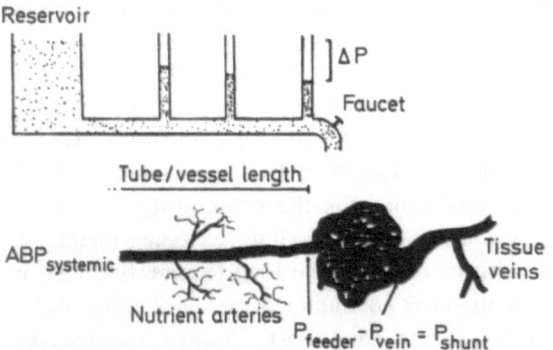

Fig. 4. Nornes' model (1980) illustrates the pressure conditions in vessels leaving the angioma feeder. With the faucet closed, all vertical tubes show the same pressures. With the faucet open, as shown in the illustration, pressure in the vertical tubes decreases with distance to the reservoir

It is difficult to describe physically flow patterns in a constantly changing vascular bed with many branches, elastic walls and unsteady pulsatile flow. Nevertheless, Nornes' model shows that with increasing length of a vessel, the pressure in its branches drops. Applied to angioma conditions, this means that with increasing length of the feeding artery, the Δp (pressure gradient) in branching arterioles decreases. As these branches supply adjacent brain, this has major clinical implications.

4.2. Hemodynamics in the Angioma

Contrary to the normal situation, blood flow through angiomas is largely pressure dependent. The characteristic flow-pressure relation of such passive vascular systems is shown in Fig. 5.

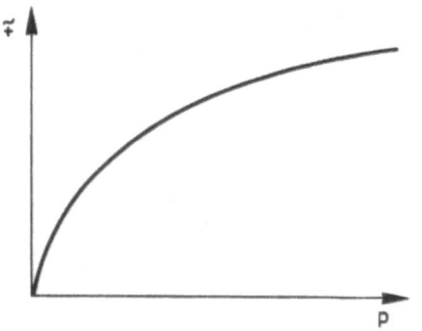

Fig. 5. Flow pressure relation in passive vascular systems (F = flow rate, P = pressure)

With rising pressure, the corresponding increase of flow rate levels off (Green 1963, McDonald 1974).

Feeding arteries show high flow velocities. Very high velocities are said to result in turbulent flow (Nornes 1980) and thereby increasing flow resistance. Depending on where the turbulence occurs, different feeders may have different resistances in one and the same angioma.

4.3. Brain Perfusion and Angioma

Brain perfusion in angioma patients is impaired by low arterial inflow pressure and high venous outflow pressure. Angiographically, this corresponds to poor visualization of brain regions adjacent to the AVM. The distribution of blood flow is impaired, the veins are dilated. In angiography, sometimes even retrograde venous flow can be demonstrated.

The extent to which perfusion of the adjacent brain is reduced depends on its distance to the AVM and on the pressure gradient (Δp) in the angioma. Nothing is known about the blood flow velocities in brain supplying vessels, about autoregulative capacity and actual widths of the regulating, brain supplying arterioles.

The lower limit of vascular autoregulation is the blood pressure under which the flow rate starts to drop (Harper 1965, Lassen 1964). Reaching this limit, the arterioles become maximally dilated (Fig. 6). This is reported to happen at 50–60 mmHg mean arterial pressure in man, and at a somewhat higher rate in people with hypertension (Harper 1965, Nornes 1977 and Strandgaard 1973). It can be assumed that under pathologically low pressure conditions as found in angiomas, arterioles are dilated.

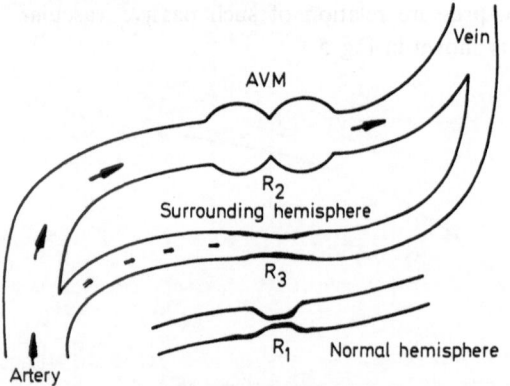

Fig. 6. The stream resistance of normal arterioles ($R 1$) is much higher than that of the AVM ($R 2$). Brain supplying arterioles in the AVM's surroundings have to dilate ($R 3$) in order to ensure tissue perfusion (Spetzler 1978)

Lowered perfusion pressure and steal phenomena are reported to reduce the perfusion of AVM surrounding brain and thereby cause focal neurological signs and psychosyndromes. As a result of the autoregulative process, angioma patients can, however, remain free of clinical symptoms. Computertomographic and magnetic resonance studies often did not suggest local hypotrophic changes in such patients.

In summary, the hemodynamic situation of the brain in angioma carriers is therefore characterized by
1. Low arterial pressure.
2. High venous pressure.
3. Impaired blood distribution.
4. Steal phenomena.
5. Underperfusion of adjacent brain?
6. Impaired autoregulation?

4.4. Hemodynamic Changes After Angioma Exclusion

Angioma surgery is often assumed to be beneficial, especially in those patients who show obvious steal phenomena. However, intraoperative and postoperative experience does not always confirm this, because swelling, hyperemia and bleeding may be encountered (Mullan 1979, Nornes 1979 and Spetzler 1979). Impaired autoregulation is reported to play a major role in these cases.

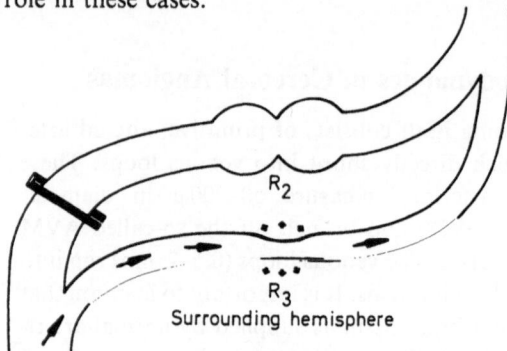

Fig. 7. With AVM occlusion, $R 2$ becomes much higher than $R 3$. The whole blood flow is now directed into brain-supplying vessels which have to withstand the now "normal perfusion pressure" after long-standing previous dilation. In the case of an impaired autoregulation, the intracapillary pressure would rise far above normal values and probably result in extravasation and/or rupture (Spetzler 1978)

What happens if maximally dilated arterioles are unable to react to the rising perfusion pressure after AVM removal? Experimental findings suggest that vasomotor regulation is impaired after long-standing dilatation. Folkow (Folkow 1968, 1971) found structural changes in arterioles following chronic hypotension. Muscle fibers in the tunica media were rarified. The vessels were enlarged and showed poor contractility. If this applies for the angioma patient, severe hyperperfusion or circulatory "breakthrough" would be the consequence of angioma exclusion (Eckström-Jodal 1971, Spetzler 1978, Nornes 1979).

Spetzler experimentally propounded his "normal perfusion pressure breakthrough theory". Nornes thought that the length of feeding arteries is of additional significance for the breakthrough phenomenon. He found brain swelling to occur mainly in AVM with more than 8 cm long feeders. Swelling or edema were never encountered in over 80 angiomas if the feeding arteries were short. These even applied for high flow AVM (Nornes 1980).

II. Animal Experiments

1. Experimental Arteriovenous Fistulae

Spetzler (1978) and Scott (1978) were the first to induce experimental angioma-like conditions in cats (Spetzler) and apes (Scott). Based on his experimental findings, Spetzler described the factors that he thought to be relevant for complications after AVM surgery ("normal perfusion pressure breakthrough theory").

Both authors used extracranial AV fistulas, expected them to reduce cerebral perfusion pressure and thereby imitate the effects occurring in cerebral angiomas. True cerebral fistulas have never been applied in animals.

Our attempts to develop such an intracranial fistula model in dogs failed, as all artificial cerebral shunts obliterated spontaneously within 10–20 minutes. This was the reason why we were forced to also use cervical AV fistulas. Nevertheless, we did try to develop animal models that more closely resemble cerebral angiomas than the models described in the literature.

The cerebral vascular anatomy in cats differs considerably from that in man (Nickel 1984). Most of the brain's blood supply comes from the maxillar artery. This vessel issues from the common carotid artery and forms a rete mirabile in the area of the foramen ovale which is connected to the Circle of Willis by rami retis (Newton 1974). The cerebral arterial supply in apes and rats resembles more the human situation (Green 1968, Scott 1978, Bannister 1984). The rat brain's venous drainage shows a special characteristic: The blood in the transverse sinus leaves the cranium via the internal maxillar, posterior facial and then jugular vein (Green 1968). Cervical AV fistulae therefore increase the intracranial venous pressure.

1.1. The Fistula Models of Spetzler and of Scott

The stream resistance of an AV fistula is low. Thus, most of the flow volume of its feeding artery is directed into the shunt. Brain regions being supplied by branches of these feeders therefore have abnormally low perfusion pressure. Chronic dilation of arterioles in these areas is thought to suspend their adaptability to pressure changes. Following exclusion of the fistula, perfusion pressure of the surrounding brain rises to normal levels. According to Spetzler's theory, this pressure now "breaks through" into the capillaries, thus leading to rupture (hemorrhage) and plasma extravasation (edema).

His theory is based upon the assumption of impaired autoregulation of brain supplying branches of the angioma feeders. In 30 cats, Spetzler applied end-to-end anastomoses between the distal common carotid artery (CCA) and the proximal common internal jugular vein (VJC). As shown in Fig. 8 a, the feeding artery of this extracranial fistula was the contralateral CCA that also supplied the brain. Shunt flow was electromagnetically measured on ipsilateral distal and contralateral CCA. Autoregulation was studied directly after the application of the fistula and 6 weeks later, directly following its occlusion.

As a working model for angiomas, there are some shortcomings in this experiment:
— The highest pressure drop occurs far more distal from where brain supplying vessels branch off.
— There is no pressure rise in the brain's venous system.
— Pressure and flow of the brain arteries cannot be measured, cerebral hemodynamic changes therefore have to be omitted. Scott's model was somewhat more suitable (Fig. 8 b). He applied side-to-side anastomoses, so that the jugular vein had high pressure cranial flow. However, other shortcomings of the Spetzler model persisted.

1.2. Own Fistula Models

Two different types of cervical AV-fristulas have been worked out in rats and later in cats:

a) Direct fistula (T-fistula)

In the T-fistula, end-to-side anastomosis between the proximal CCA and the VJC is applied. Flow velocities

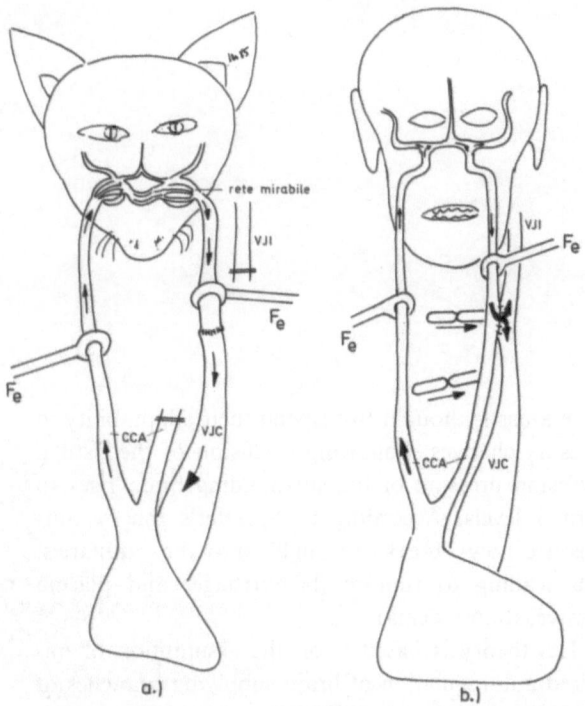

Fig. 8. Schematic drawings of Spetzler's animal model in cats (a) and Scott's model in apes (b). (*CCA* A. carotis communis, *VJC* V. jugularis communis, *VJI* V. jugularis interna, *Fe* electromagnetic measuring probe)

and flow rates of AVMs and also their impacts on vessel walls are thereby accurately imitated (Fig. 9 a).

b) Indirect fistula (H-fistula)

In the H-fistula, the external jugular vein was cut distally and connected end-to-side to the CCA (Fig. 9 b and Fig. 10). In this model, the proximal CCA is the shunt feeding artery, whereas the distal CCA represents a brain supplying artery with low intravascular pressure. The fistula fills the veins with arterialized blood. Cranial venous flow occurs, leading to elevated intracranial venous pressure.

Other than in Spetzler's or Scott's model, autoregulation of brain supplying arteries leaving the feeder distal to the shunt could be studied. The distal CCA shows low pressure and in our model represents a brain supplying artery distal from the AVM. According to Spetzler's theory, this should be the vessel in which the pressure breaks through after fistula occlusion.

1.3. Topics of Our Experimental Investigations

1. How does the flow velocity and profile change with high flow in fistulas?

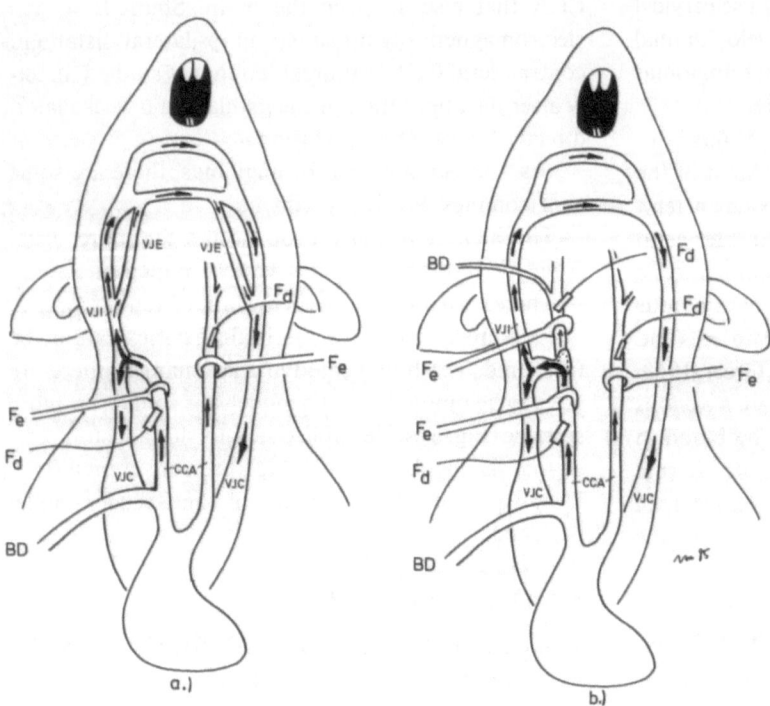

Fig. 9. (a) Direct fistula between CCA and jugular vein (T-fistula). (b) Indirect fistula where the external jugular vein is connected end-to-side to the CCA (*CCA* A. carotis communis, *VJC* V. jugularis externa, *VJI* V. jugularis interna, *BD* blood pressure, *Fe* electromagnetic probe, *Fd* Doppler probe, *ST* transverse sinus, *VFA* V. facialis anterior)

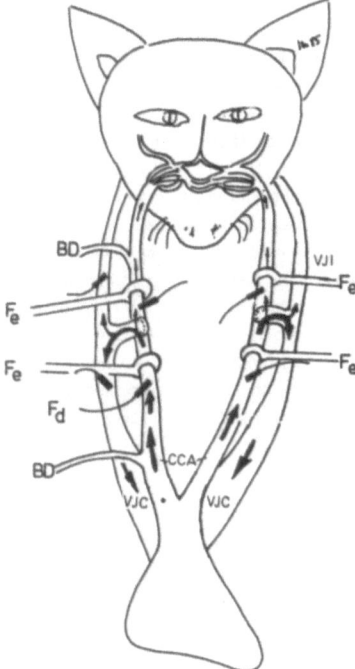

Fig. 10. Bilateral H-fistula in cat

2. How does the flow rate change in acute and chronic (duration 8 weeks or more) fistulas?

3. How does vascular resistance change with time?

4. How far does the resistance index (according to Pourcelot) drop in direct and indirect fistulas?

5. Are there differences in flow rates and stream resistance between purely fistula feeding arteries (T-fistulas) and arteries that supply both fistula and the brain (H-fistula)?

6. Electromagnetic flow measurements and intra-operative Doppler sonography had to be compared.

7. Do cervical AV-fistulas lead to impaired vascular autoregulation in the brain?

8. Is Spetzler's theory confirmed in different experimental settings?

9. What is the impact of an AV-fistula on systemic circulation?

2. Materials and Methods

2.1. Anaesthesia

a) Rats

Surgery and investigations were carried out under neuroleptic analgesia. Depending on the duration, 0.5–1.2 ml of Thalamonal✓ were given. All animals showed sufficient spontaneous respiration. Arterial pO_2 and pCO_2 were randomly checked and found to be within normal limits.

b) Cats

The cats had phenobarbitol analgesia (25 mg/kg). In addition, they received 1 ml atropin intraperitoneally. Pancuronium✓ (0.06 mg/kilogram) was used for relaxation. Artificial respiration was applied with the Hugo Sachs electronic device according to Schuler.

2.2. Surgical Procedure

a) Rats

The rats were placed on an X-ray permeable pane in a supine position. All anastomoses were applied on the right CCA because this vessel is most appropriately visualized by retrograde brachial angiography. The right brachial artery was used for intraoperative blood pressure monitoring and angiography.

Following a mid-line incision, the cervical part of the right CCA was carefully exposed from the clavicula to the vessel's bifurcation. After dissection of cervical veins, anastomoses could then be sewn without vessel distension. Fistula applications can be seen in the figures 9 and 10. In the T-fistula, the proximal CCA was connected end-to-side to the common jugular vein. In the H-fistula, orthograde arterial flow was preserved with end-to-side anastomosis of the external jugular vein to the CCA. After clipping of these vessels, anastomoses were sewn with single button sutures. Patency of the fistula was immediately demonstrated by enlarging red veins. In every case, fistula flow was documented dopplersonographically, electromagnetically and angiographically. The same investigations were repeated after two and three months.

b) Cats

Twenty-two cats underwent the same surgical procedure. Ten additional cats received the application of a bilateral H-fistula. Four of these cats were trephined in the right precentral area in order to study the microcirculation (laser Doppler) and flow velocities (conventional Doppler) in cortical vessels before and after temporary fistula occlusion.

2.3. Blood Pressure Measurements

Systemic arterial blood pressure (SABP) was measured in the brachial and the distal external carotid artery. We used Statham P 23 ID transducers in connection with a Hellige ICU monitoring device.

2.4. Electromagnetic Measurements of Flow Rate

As opposed to regional or global CBF measurements (Kety 1948, Lassen 1961), electromagnetic techniques detect the flow rates of

$$R = \frac{S-D}{S}$$

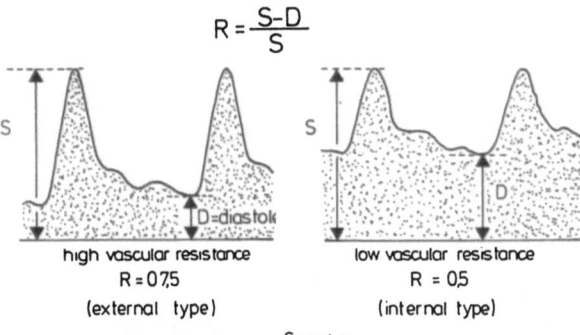

Fig. 11. Resistance index (R) as a measure of peripheral stream resistance (Pourcelot 1974)

single vessels. Some probes allow recordings in very small vessels of 0.5 mm diameter. Zero adjustment to the base line of the device was achieved by temporary clipping. Using the flow patterns thus obtained, peripheral stream resistance can be assessed mathematically. According to Pourcelot (1974), the resistance index R is a measure of the peripheral stream resistance (Fig. 11).

In general, low vascular resistance is characterized by high diastolic flow velocities (internal type), whereas low diastolic flow indicates high resistance (external type).

2.5. Dopplersonographic Measurements of Flow Velocity

Kalmus (1954) and Baldes (1958) were the first to use ultrasound for measuring the flow velocity in blood vessels. In 1959 Satomura published his studies on dopplersonographically recorded flow patterns. Since then, measurements in increasingly smaller vessels have become possible.

The Doppler effect (Christian Doppler, Vienna, 1803–1853) lies in the occurring frequency shift when transmitter and receiver of any kind of physical wave show a relative movement (Doppler 1842). This also occurs when transmitter and receiver are fixed, but the reflector is mobile.

In Doppler measurements of blood flow, erythrocytes serve as reflectors of ultrasound waves being transmitted and received by the Doppler probe. The frequency shift caused by moving red cells is called "Doppler frequency" or Doppler shift.

$$f = \frac{2 \, f o \cdot v \cdot \cos \alpha}{\gamma}$$

v = blood flow velocity (cm/sec), f_0 = mean transmitter frequency (Hz), α = angle between ultrasonic beam and flow direction, γ = ultrasound velocity in tissue (cm/sec)
f = Doppler shift (Hz)

According to this formula, the Doppler frequency is proportional to blood flow velocity.

2.6. Measurements of Microcirculation (Laser Doppler)

The laser Doppler facilitates direct investigation of regional circulation. This method was initially used by Riva (1972) to study retinal perfusion in rabbits. The first measurements in human retinas were performed by Tanaka (1974). Further technical progress (Nilsson 1980, Gygax 1983) led to easy applicability in experimental and clinical situations.

We used a laser Doppler device (Periflux PF 101) with a heliumneon laser source of 2 mW. Its monochromatic light is transduced into the measuring probe and emerges into the tissue, where it is dispersed and scattered back. Reflection from fixed objects leaves the frequency unchanged whereas reflection from moving particles causes a Doppler shift. This results in the recording of a whole spectrum of different shifts which are picked up by photodetectors. The obtained signals correspond strictly to the flow velocity. All directions of flow are recorded within a radius of 1.5 mm in front of the measuring probe. Therefore, the method supplies relative values of microcirculation in capillaries and mainly serves to detect changes in microcirculation.

2.7. Angiography

Patency of the fistulas was checked by right-sided retrograde brachial angiography immediately after surgery. Further angiograms were carried out two and three months later. 1.5 ml of Solutrast[B801v002n] were injected under pressure.

3. Results of the Fistula Models in Rats

3.1. Direct AV Fistula (T-fistula)

3.1.1. Electromagnetic Flow Measurements

In twelve experimental animals, end-to-side fistulas were applied between the proximal right common carotid artery (CCA) and the right common jugular vein (VJC) (Fig. 9 a). Flow rate of the right proximal CCA was electromagnetically measured before and after surgery and two months later. After two months, two fistulas had obliterated, one animal had died.

a) Acute fistula

The mean flow rate of the CCA prior to the operation was 4.75 ml/min. It increased postoperatively to

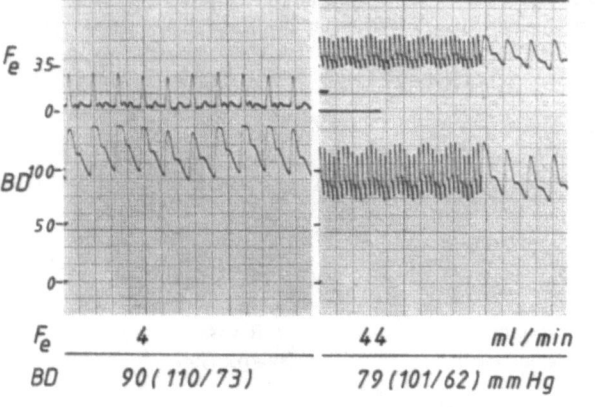
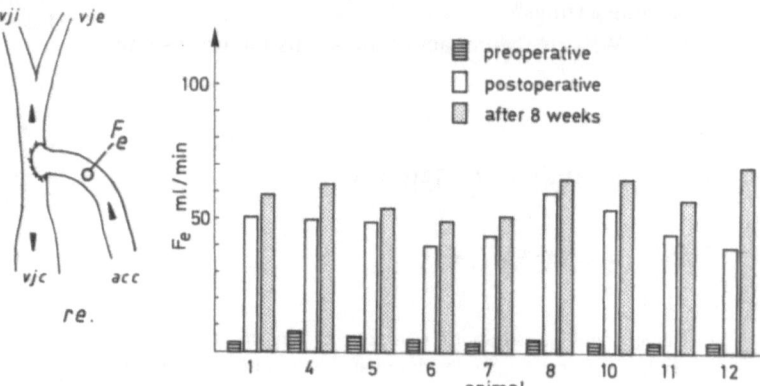

Fig. 12. Electromagnetic flow rates of right CCA (*Fe*) and systemic blood pressure (*BD*) in T-fistula; preoperative (left) and postoperative (right) recordings

46.7 ml/min (range: 33–60 ml/min) which is 9.8 times as much as the preoperative value (Fig. 12).

Preoperative flow patterns showed a high systolic peak followed by a low "catacrotic shoulder" and low diastolic values (external type). This changed following the fistula application. Due to decreased vascular resistance, diastolic flow velocities increased more than the systolic values. The resistance index of the CCA fell from a preoperative mean value of 0.87 (0.93–0.79) to 0.37 (0.28–0.46). This represents a 42% reduction of peripheral stream resistance (Fig. 13).

The left CCA showed no change in flow rate or flow velocity.

b) Chronic fistula

After two months, flow rates of the fistula feeding CCA had increased further to mean values of 59.2 ml/min (49–70 ml/min). This is 12.5 times the preoperative rate and 1.3 times the immediate postoperative value (Figs. 12, 14).

Corresponding to this, the difference between systolic and diastolic flow velocities decreased further, indicating reduced vascular resistance. The mean resistance index of the right CCA after two months was

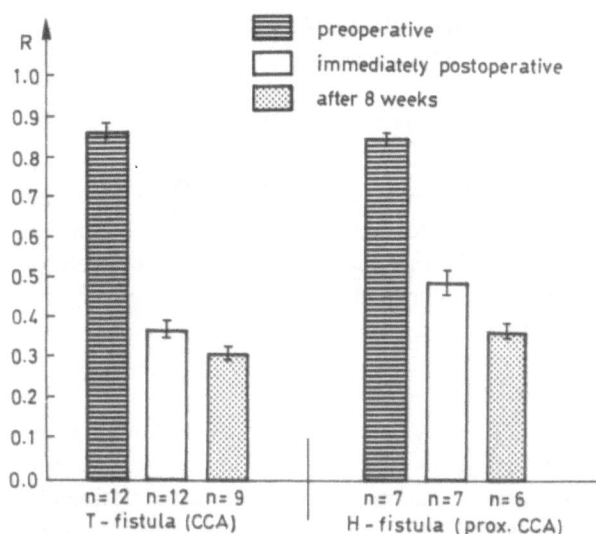

Fig. 13. Comparison of resistance indices in T- und H-fistulas under chronic and acute conditions: drop is more pronounced in T-fistulae

0.31 (0.28–0.36). Its postoperative course is shown in Fig. 13. In summary, chronic T-fistulas showed a 60% drop in stream resistance and tenfold blood flow rates compared to the normal CCA.

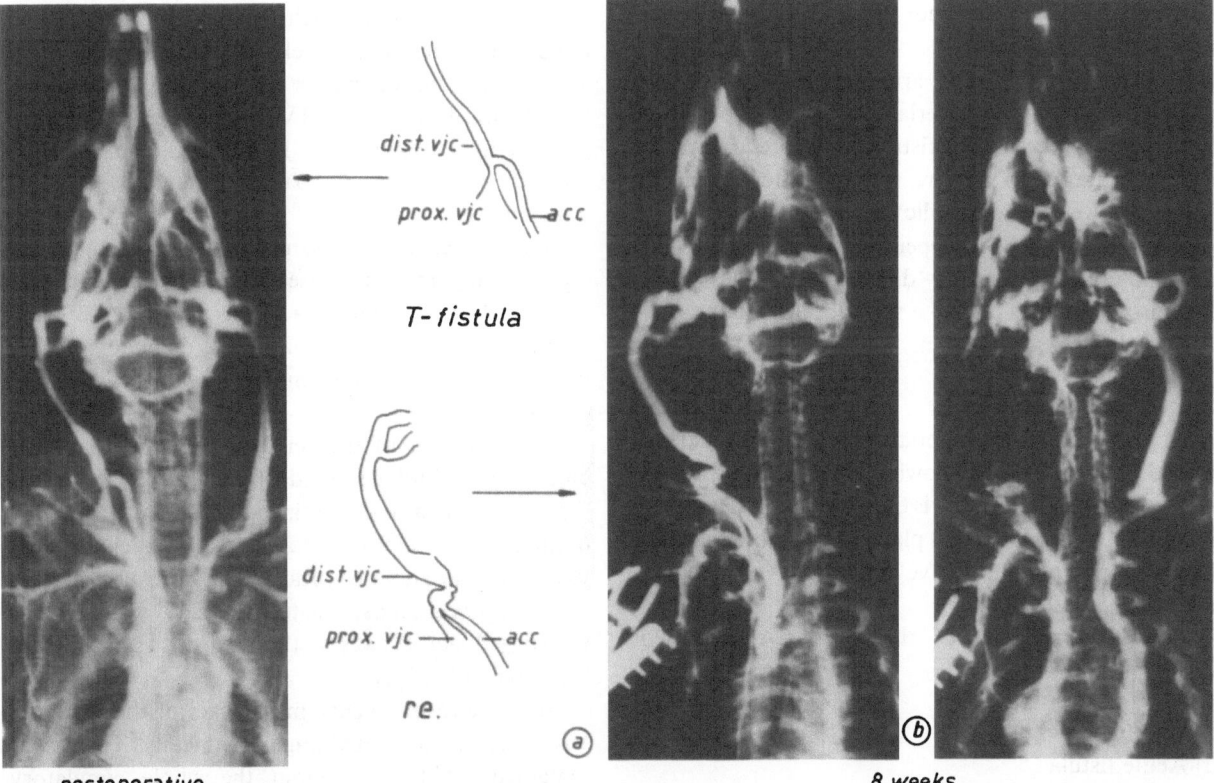

Fig. 14. T-fistula of the same rat: immediate postoperative and 8 weeks postoperative angiogram; the right side shows early (a) and late (b) picture from the serial angiography

The contralateral CCA did not show any changes in flow rate or vascular resistance, even after two months.

3.1.2. Intraoperative Doppler Sonography

Frequency spectra of the CCA and cervical veins were studied with a 20 MHz pulsed Doppler system (MF 20 Microvascular Doppler, EME, D-7770 Ueberlingen, FRG) using 2 and 3 mm sized Doppler probes before, immediately after and 8 weeks following surgery. Contralateral vessels were also investigated.

a) Acute fistula

Preoperative flow velocity of the CCA showed steep systolic upstrokes, high peaks and low diastolic values. After surgery, diastolic velocities increased noticeably and even exceeded the upper limits of the device's recording range. In all cases, this resulted in a non-pulsatile band-like frequency spectrum up to 12.5 kHz. Frequencies above this were not displayed. This could be due either to the filter setting in the Doppler device or to its pulse repetition frequency of 25 kHz (Cathignol 1983). Frequencies above 12.5 kHz, which is exactly half of the pulse repetition frequency, are not recorded. This upper frequency limit corresponds to a flow velocity of 84 cm/sec. As velocities in T-fistulas reached 105 cm/sec, they were not detected by Doppler sonography.

Frequency spectra of arterialized veins showed mixed flow patterns of arterial and venous type. With increasing distance to the fistula, the more artery-like curve changed into a more venous pattern. Venous spectra showed a slow systolic increase, a more rounded profile and respiration-dependent deflections. Also, contralateral venous flow differed from the norm. Increased velocities could be seen dopplersonographically and on the angiogram.

b) Chronic fistula

After two months, the arterialized veins had enlarged considerably while still showing the same flow velocities. As the blood flow rate is the mathematical product of a vessel's cross-sectional plane and its flow velocity, shunt flow rates must have noticeably increased in chronic fistulas.

Frequency spectra of the left CCA remained unchanged.

3.1.3. Blood Pressure

a) Acute fistula

The preoperative mean value of systemic arterial mean blood pressure was 94.3 mmHg (80–112 mmHg). After fistula application it immediately dropped to 83.7 mmHg (74–103 mmHg), which is an 11.2% decrease.

b) Chronic fistula

After two months, arterial mean blood pressure had risen to almost the preoperative mean values of 89.7 mmHg (84–100 mmHg).

c) Fistula occlusion test

Temporary clipping of the fistula was performed in order to study its hemodynamic significance for the systemic circulation.

— Acute fistula

Occlusion of acute fistulas led to a 16% increase in systemic arterial mean blood pressure (Fig. 15).

— Chronic fistula

Eight weeks after fistula application, occlusion resulted in a 20% rise of arterial mean blood pressure.

3.1.4. Angiography

Each animal had a brachial arteriography before, immediately after and 8 weeks following surgery. In all angiograms, the diameters of the CCA were determined. Distal and proximal VJC were assessable only in postoperative angiograms. In all cases, the points of measurement were in 5 mm distance from the anastomosis.

The immediate postoperative CCA diameters increased only in some cases. Nevertheless, this was statistically significant (p = 0.05).

In the following two months, the diameters further increased significantly from a mean 1.03 to 1.22 mm (p = 0.01).

The diameter of the veins doubled, so that their cross-sectional plane quadrupled. Maximal venous diameters were also determined and found to rise up to 6.2 mm. The intracranial transverse sinus enlarged as well as contralateral cervical veins.

After 8 weeks most anastomoses showed stenoses at the surgial site. However, they did not seem to have hemodynamic effects.

In some cases, aneurysm-like sacs of the proximal VJC were demonstrated angiographically. They often remained visible throughout the whole circulatory phase. Months later, these sacs were operatively exposed and found to have extremely thin walls.

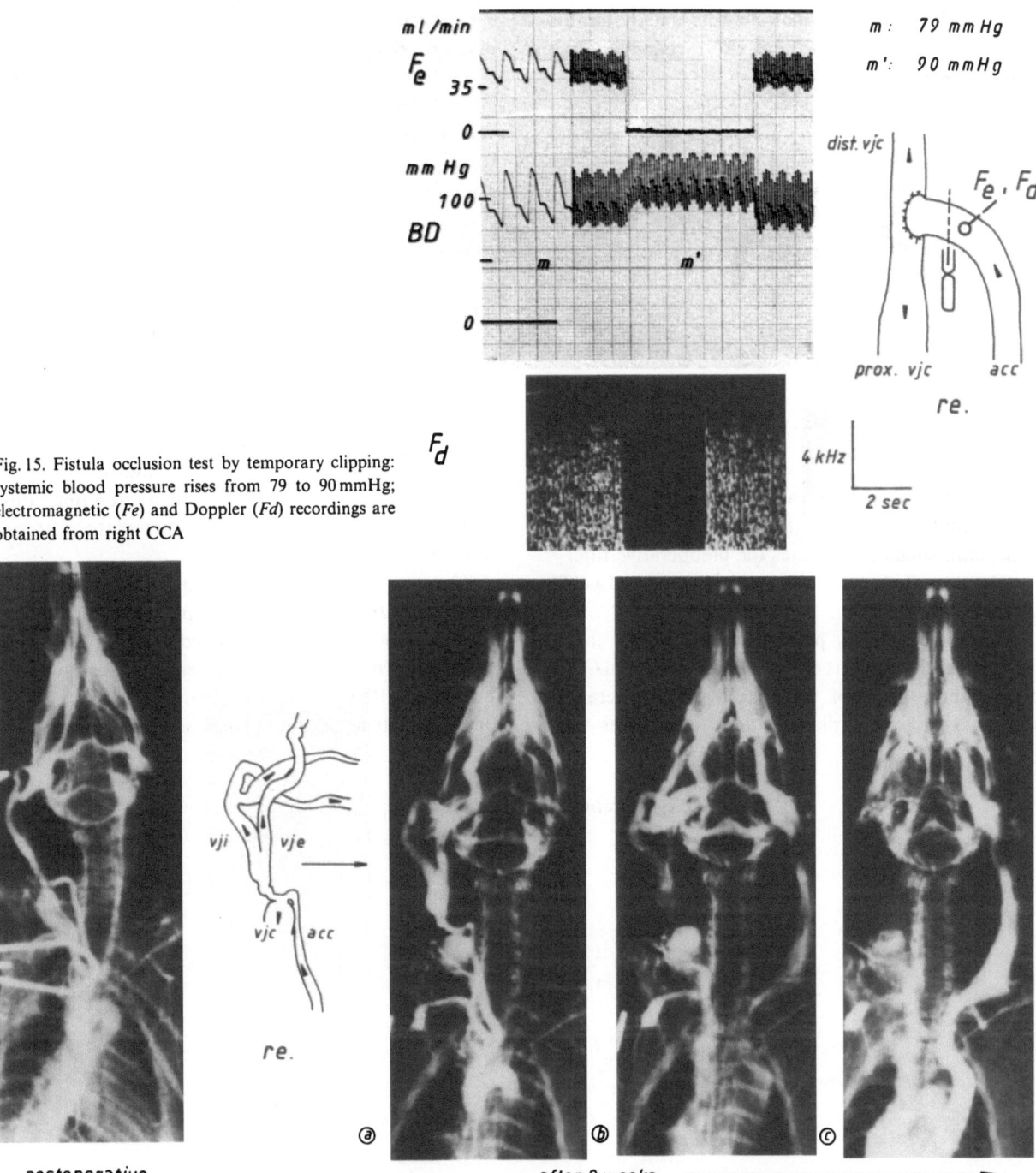

Fig. 15. Fistula occlusion test by temporary clipping: systemic blood pressure rises from 79 to 90 mmHg; electromagnetic (*Fe*) and Doppler (*Fd*) recordings are obtained from right CCA

Fig. 16. Comparison of angiograms taken immediately and 8 weeks after T-fistula application. On the right side, 3 different phases of the serial angiography (a, b, c) show an aneurysmatic sac of the right proximal VJC

3.2. Indirect AV Fistula (H-fistula)

In seven animals, the distal end of the right external jugular vein (VJE) was connected to the side of the right CCA (Fig. 9 b). Before and after surgery, electromagnetic flow measurements and Doppler sonography were carried out. Arterial blood pressure was con-tinuously measured in the right brachial artery and also in the right CCA (catheter insertion via external carotid artery). All animals were reinvestigated two months later. In three animals, the left CCA was occluded after 8 weeks in order to induce further hypoperfusion of the brain. They were kept for additional studies after 12 and 18 weeks. In one animal, the fistula had obliterated after two months while the CCA remained patent without stenosis.

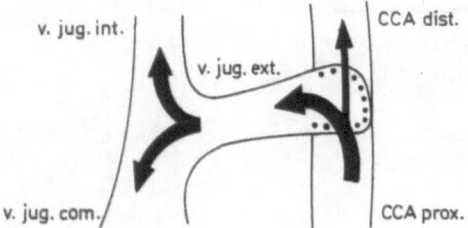

Fig. 17. Flow directions in the H-fistula

3.2.1. Electromagnetic Measurements

a) Acute fistula

Preoperative mean flow rate of the right CCA was 4.4 ml/min (3–6 ml/min). After surgery, it increased to 33 ml/min, which is 7.6 times the preoperative rate.

The mean flow rate of the distal CCA showed a 60% decrease to 2.4 ml/min.

Postoperative flow patterns were different in the proximal and distal part of the CCA. In the distal CCA, only slight changes of the normal pattern were observed. Systolic and diastolic flow amplitudes were

proportionately reduced. The mean pressure in this vessels dropped to 47 ± 5 mmHg (Fig. 18).

The proximal CCA showed postoperative changes similar to those in T-fistulas. Systolic and diastolic flow rates increased significantly. Characteristically, increase in the diastolic values was considerably more pronounced due to low peripheral resistance. The resistance index was calculated according to Fig. 11 and found to drop by 50% from preoperative mean values of 0.85 (0.80–0.88) to postoperative 0.49. In the distal CCA, this index decreased only slightly to 0.81.

Postoperative measurements at the contralateral CCA did not reveal any changes.

b) Chronic fistula

Eight weeks after surgery, the flow rate of the right proximal CCA was 47.7 ml/min, which is ten times the normal value and 1.4 times the immediate postoperative rate (Fig. 19).

The mean flow rate of the distal CCA did not show further significant changes compared to the acute fistula. After 8 weeks, it reached 70% of the preoperative normal value. The difference between systolic and diastolic flow rates further diminished, which again corresponds to an additional increase in mean flow rate

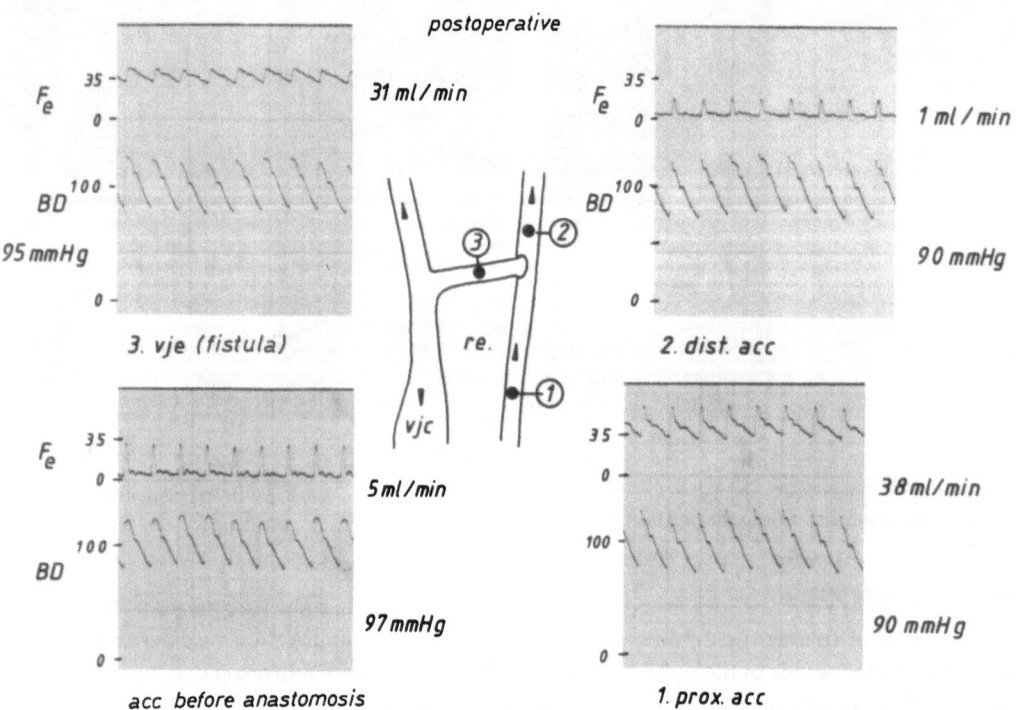

Fig. 18. Electromagnetic recordings (*Fe*) in CCA before and immediately after H-fistula application. Flow rates and profiles are very different in the proximal and distal CCA following surgery (acc = CCA)

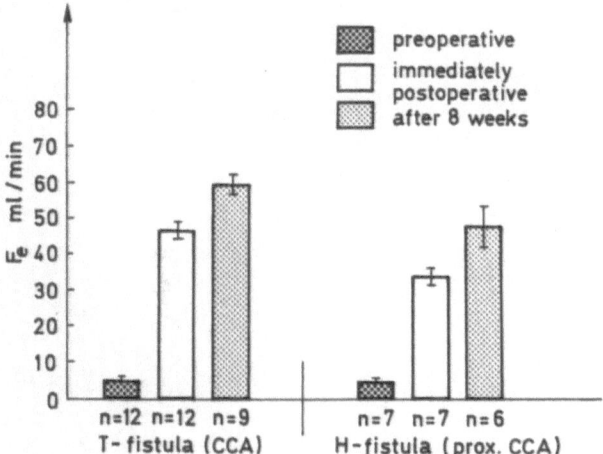

Fig. 19. Electromagnetic flow rates (*Fe*) in the proximal CCA; comparison of T- and H-fistula

of the vessel. This simply means that the fistulas enlarged in the postoperative period.

The resistance index of the proximal CCA had dropped further to 0.37 (0.28–0.45) after two months.

Resistance in the distal CCA after two months of H-fistula was 0.82 and thereby almost maintained pre-operative values.

Resistance indices of proximal and distal CCA diverge considerably immediately following the application of an H-fistula, whereas in the ensuing period only minimal changes occur (Fig. 20). Again, the contralateral CCA showed no hemodynamical changes.

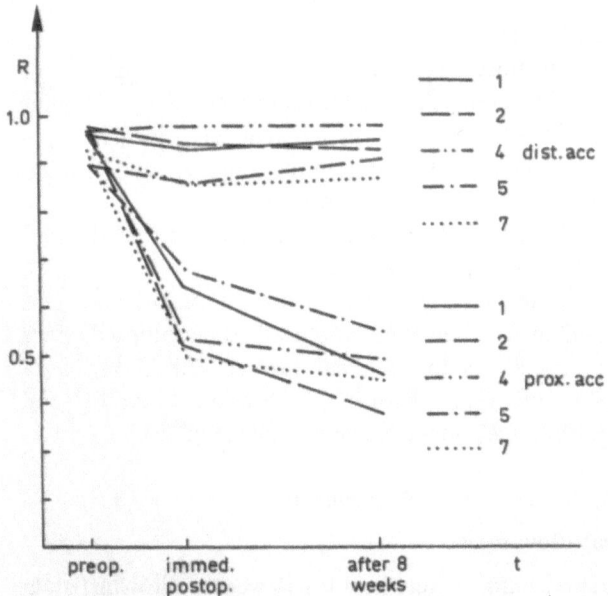

Fig. 20. Pre- and postoperative resistance indices of proximal and distal CCA in H-fistulae (animals *1, 2, 4, 5, 7*)

Comparison of resistance indices of the proximal CCA in T-fistulas and H-fistulas

The resistance drop of fistula feeding, proximal CCA was more pronounced in the T-fistula than in the H-fistula.

3.2.2. Doppler Sonography

a) Acute fistula

The proximal CCA and the fistula itself gave a non-pulsatile, band-like, postoperative frequency pattern. The upper recording limit of the device being 12.5 kHz, higher frequencies again were not measurable. The resulting upper recording limits for a 1 mm thick CCA were 84 cm/sec for flow velocity and 39.6 ml/min for flow rate.

As a consequence, fistula flow could only be investigated when it was below 30 ml/min. Under this condition, dopplersonographic flow patterns were similar to electromagnetic recordings. Flow velocities in distal right and left CCA were normal.

The veins again provided mixed patterns of the arterial and venous type. The internal jugular vein showed a high systolic peak, a high diastolic flow and respiration-dependent deflections. The flow pattern of the right VJC was the same as in the fistula itself due to low resistance.

b) Chronic fistulas

Due to further increase in flow rate and velocity, all Doppler measurements in the chronic H-fistula exceeded the recording range of our device.

3.2.3. Blood Pressure Measurements

a) Acute fistula

Changes of mean arterial blood pressure (BP) were similar to that in T-fistulas. After application of the H-anastomosis, pressure values dropped from pre-operative 92.9 mmHg to 81.9 mmHg in the systemical circulation and to 47 ± 5 mmHg in the distal CCA.

b) Chronic fistula

After two months, the mean systemic arterial BP had reached preoperative values (92.5 mmHg), whereas pressure in the distal CCA remained stable. Thus, systemic circulatory changes were compensated within two months.

c) Fistula occlusion test (Figs. 21, 22)

The venous part of the anastomosis was temporarily occluded in order to re-establish "normal" CCA-flow.

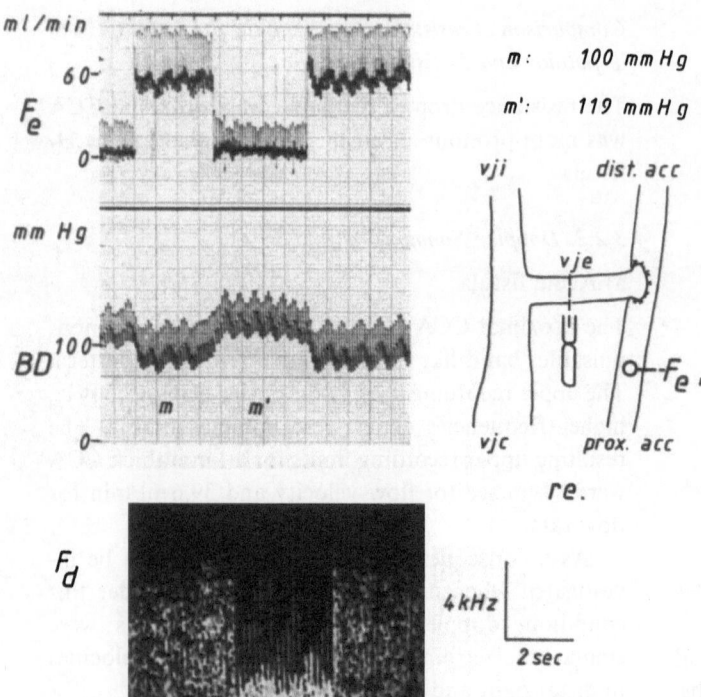

m: 100 mm Hg

m':   119 mm Hg

Fig. 21. Fistula occlusion test: Temporary clipping of the VJE results in a systemic blood pressure (*BD*) increase. Electromagnetic (*Fe*) and Doppler (*Fd*) recordings are obtained from the proximal CCA

Temporary occlusion of acute and chronic H-fistulas resulted in a 20% increase in systemic mean arterial BP. This effect appeared within one second following the occlusion and must be explained by a corresponding rise in systemic vascular resistance.

3.2.4. Angiography

Diameters of the CCA and internal jugular vein were determined before and after fistula application. The points of measurement were 3 mm distal and proximal to the anastomosis.

Immediately after surgery, mean diameters of proximal and distal CCA had not changed. Even in the following eight weeks, significant enlargement did not occur (proximal CCA: p = 0.1; distal CCA: p = 0.5).

The VJC diameter, however, doubled within eight weeks, and the internal jugular vein reached 2.5 times its initial diameter. Cross-sectional planes of these vessels consequently increased by factors of 8 (VJC) and 16 (VJI).

In most cases, stenoses of the venous part of the fistula developed. The locations werę always in some distance from the suture, so that direct surgical manipulation could not have been the cause. As in T-fistulas, aneurysm-like sacs were observed.

Intracranial transverse sinus and contralateral cervical veins were also enlarged.

In the T-fistulas, external and internal jugular veins had always been angiographically superimposed. Since in H-fistulas the external jugular vein served as a shunting vessel, proper visualization of cranial and intracranial flow became possible (Fig. 23).

3.2.5. Long-term Experiments

a) Flow rates

Three animals were kept for 18 weeks following fistula application. During this period, flow rates in proximal and distal CCA were measured four times. These

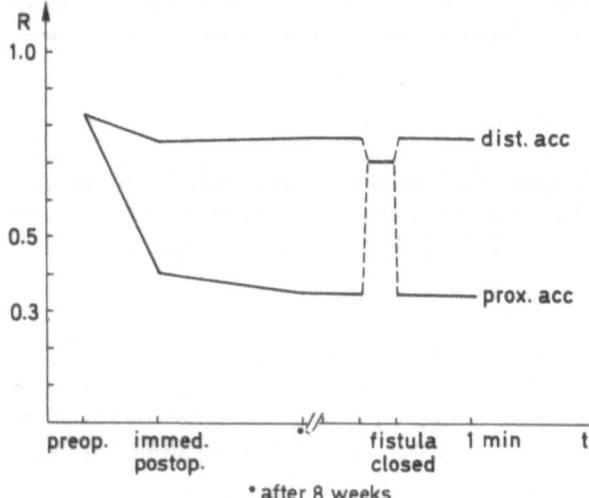

Fig. 22. Fistula occlusion and resistance index: Resistance indices (*R*) of the proximal and distal CCA divergage after H-fistula application. Occlusion of the 8-week-old fistula for a few seconds does not result in significant changes after reopening

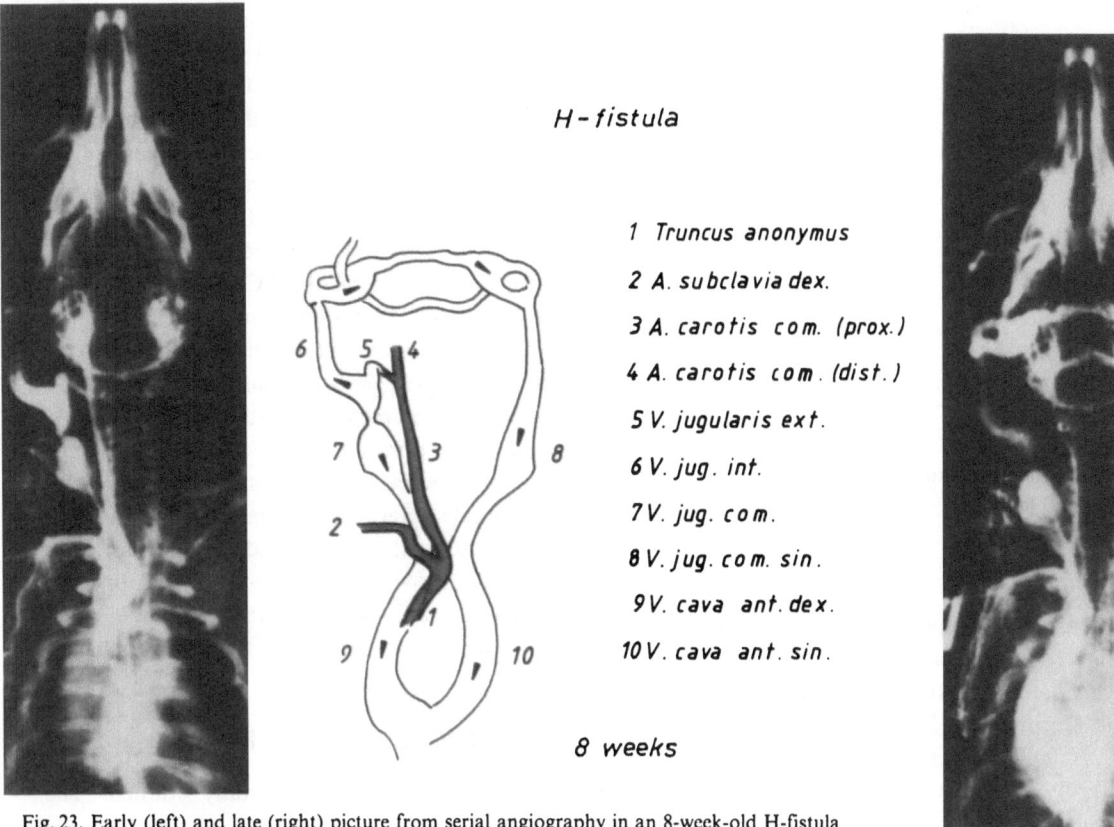

H-fistula

1 Truncus anonymus

2 A. subclavia dex.

3 A. carotis com. (prox.)

4 A. carotis com. (dist.)

5 V. jugularis ext.

6 V. jug. int.

7 V. jug. com.

8 V. jug. com. sin.

9 V. cava ant. dex.

10 V. cava ant. sin.

8 weeks

Fig. 23. Early (left) and late (right) picture from serial angiography in an 8-week-old H-fistula

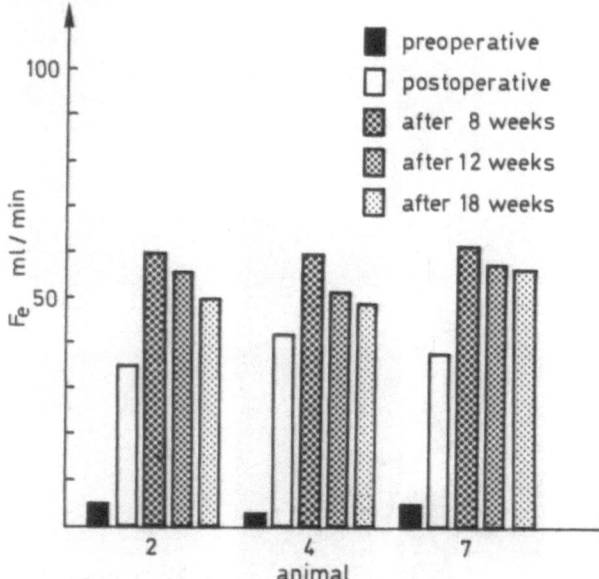

Fig. 24. Electromagnetic flow rates of the proximal CCA in H-fistula; preoperative, immediate postoperative, 8-, 12-, and 18-week postoperative recordings

animals had additional ligation of the contralateral CCA twelve weeks after the first surgery. Fig. 24 shows the significant increase of flow rates in the proximal CCA, reaching its maximum after eight weeks. The values then decrease slightly, but nevertheless stay higher than in the immediate postoperative situation. This corresponds to a slight increase of resistance indices during this period.

After 18 weeks, one animal showed exclusively cranial fistula flow (Fig. 26), but no larger increase in resistance than the others. As the VJC did obliterate in this case, resistance in the transverse sinus could not have been elevated.

Different postoperative flow resistances in proximal and distal CCA are shown in Fig. 25. Resistance drops considerably in the proximal part, reaching its minimum after eight weeks, whereas in the distal part it remains unchanged.

b) Angiography

In three animals, the vessel diameters were determined 18 weeks after H-fistula application and compared to the eight week postoperative situation. Two animals showed no change. One animal demonstrated an increase from 3.1 mm to 3.5 mm. This was due to spontaneous obliteration of the proximal VJC, so that the whole shunt volume was directed cranially through the VJI, transverse sinus and contralateral cervical veins (Fig. 26).

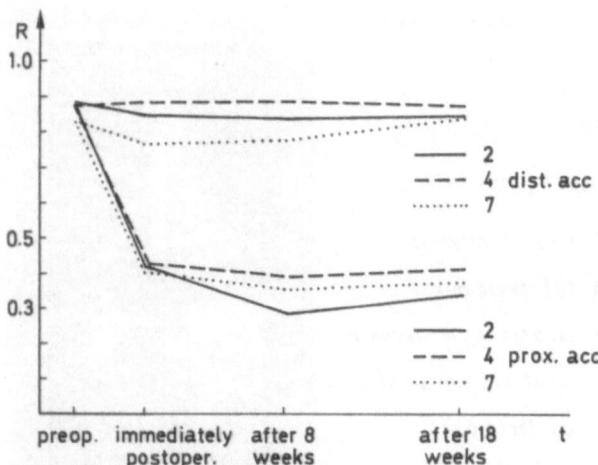

Fig. 25. Pre- and postoperative resistance indices of proximal and distal CCA in H-fistulas (animals *2, 4, 7*)

3.2.6. Autoregulation Experiments

Systemic hypotension and hypertension were medically induced in order to study hemodynamical changes in vessels supplying fistula and the brain. The drugs used were etilefrine (Effortil′, 0.5 mg i.v.) and clonidine (Catapresan′, 0.0075 mg i.v.). Flow was studied electromagnetically with simultaneous BP recording.

a) Autoregulation measured in normal CCA was normal. An almost 100% BP increase has no significant effect on flow rates in the normal CCA.

b) Flow rates in CCA proximal to H-fistula increase with rising BP and decrease with falling BP. There is no autoregulation. Flow rates are pressure-dependent.

c) Autoregulation measured in CCA distal to H-fistula was normal. An increase in BP results in slightly decreasing flow rates and amplitudes.

d) Flow rats measured in CCA proximal to H-fistula after 8 weeks remain pressure-dependent.

e) Autoregulation measured in CCA distal to H-fistula after 8 weeks shows normal autoregulation.

f) Autoregulation measurements after 18 weeks: As mentioned above, three animals had their contralateral CCA ligated 12 weeks after H-fistula application. A further six weeks later, autoregulation of CCA proximal and distal to the fistula had normalized. Even though brain inflow pressure had now dropped to 47 mmHg, the distal CCA maintained normal autoregulation with that of the proximal part still abolished.

4. Results of Fistula Models in Cats

Twenty-two cats were experimentally studied. The topics of these investigations were:

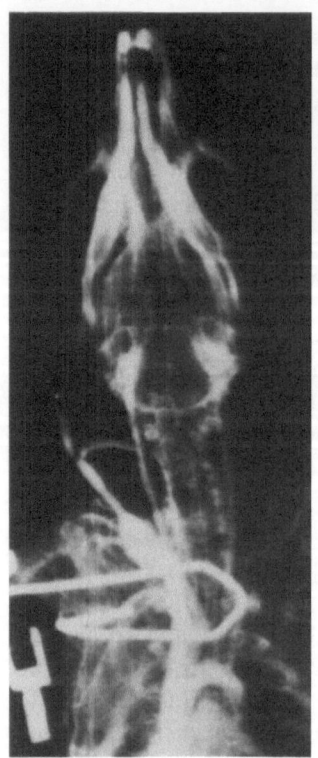
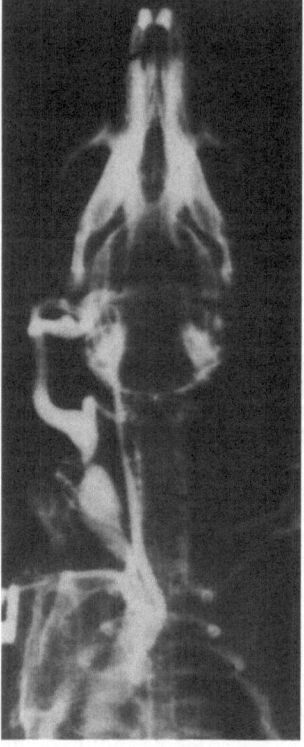
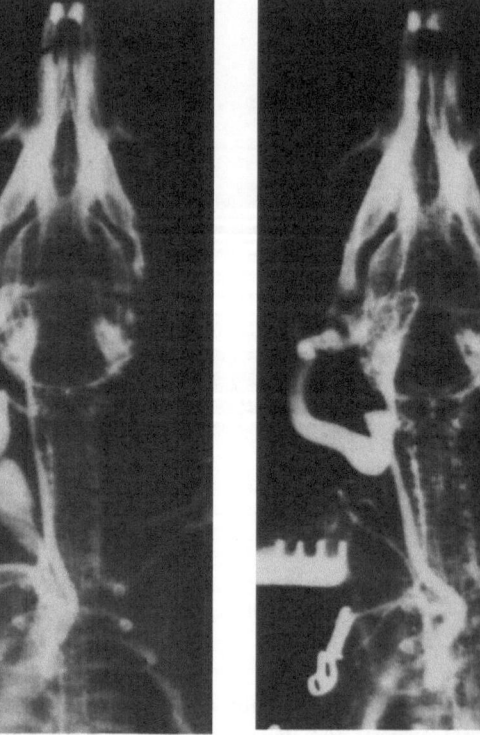

postoperative   *8 weeks*   *18 weeks*

Fig. 26. Pre- and postoperative angiograms of proximal CCA in H-fistula

—Simultaneous dopplersonographic and electromagnetical measurements in rats are impossible due to the narrow surgical approach. A direct comparison of these methods was possible in cats.

—Vibrations of vessel walls already seen in rats can be studied much better in cats.

—In order to induce more pronounced hemodynamic effects on brain circulation, bilateral H-fistulas were applied.

—Studies in Spetzler's experimental model:

a) Spetzler's fistula was reconstructed. Direct measurements of hemodynamic effects on the cortical circulation were performed (not done by Spetzler).

b) Histological studies of brains in the chronic fistula-model: Fistulas were ligated after eight weeks. Evans blue was then administered (2 ml/kg) after one hour of hypotonic, normotonic or hypertonic BP-conditions in three animals each. The cats then received 1,000 ml of a Ringer solution (8.6 g NaCl, 0.3 g KCl, 0.33 g CaCL/1,000 ml aqua dest.), followed by 800 ml formaldehyde solution (4% formaldehyde and 1% glutaraldehyde in 0.1% Sörensen buffer). The brain was then taken for histological examination (blood-brain barrier).

c) Autoregulation in acute and chronic fistulas.

4.1. Flow Rates and Blood Pressure in the H-fistula Model

The preoperative mean flow rate of the CCA was 19.2 ± 5 ml/min. After H-fistula application, the flow rate increased to 205 ml/min, which corresponds to ten-fold postoperative values already seen in rats. Flow rates of the distal CCA decreased. Even retrograde flow occurred, which had not been observed in rats. Mean values in distal CCA were 4 ml/min (—7 to +19 ml/min). Its intravascular pressure dropped to a mean 45 ± 5 mmHg.

Application of an additional, contralateral H-fistula did not have major further effects. Flow rates of proximal and distal CCA were the same as in unilateral fistulas. The pressure drop in the distal CCA was slightly more pronounced than in unilateral fistulas (43.5 mmHg). In some cases of bilateral H-fistulas, retrograde flow developed in one distal CCA (Fig. 29 b).

Eight weeks later, stenoses of the draining veins similar to those found in rats were encountered. Flow rates slightly decreased in the proximal CCA (182 ml/min) and increased in the distal CCA (6 ml/min).

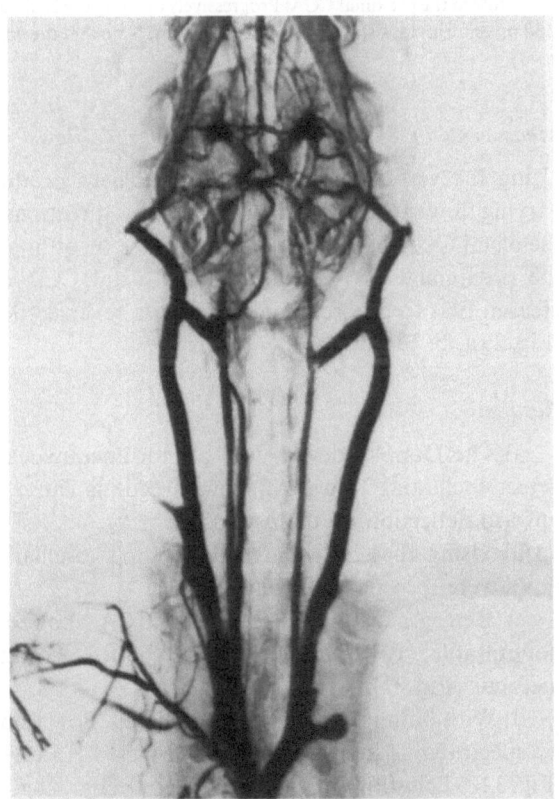 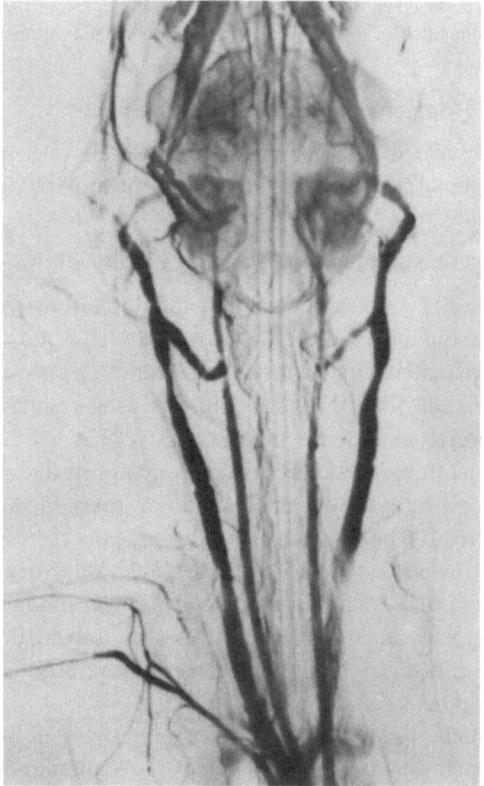

Fig. 27. Angiogram in cat with bilateral H-fistula; left: immediately following surgery, right: 8 weeks later. The 8-week angiogram shows venous stenoses as already encountered in the rat model

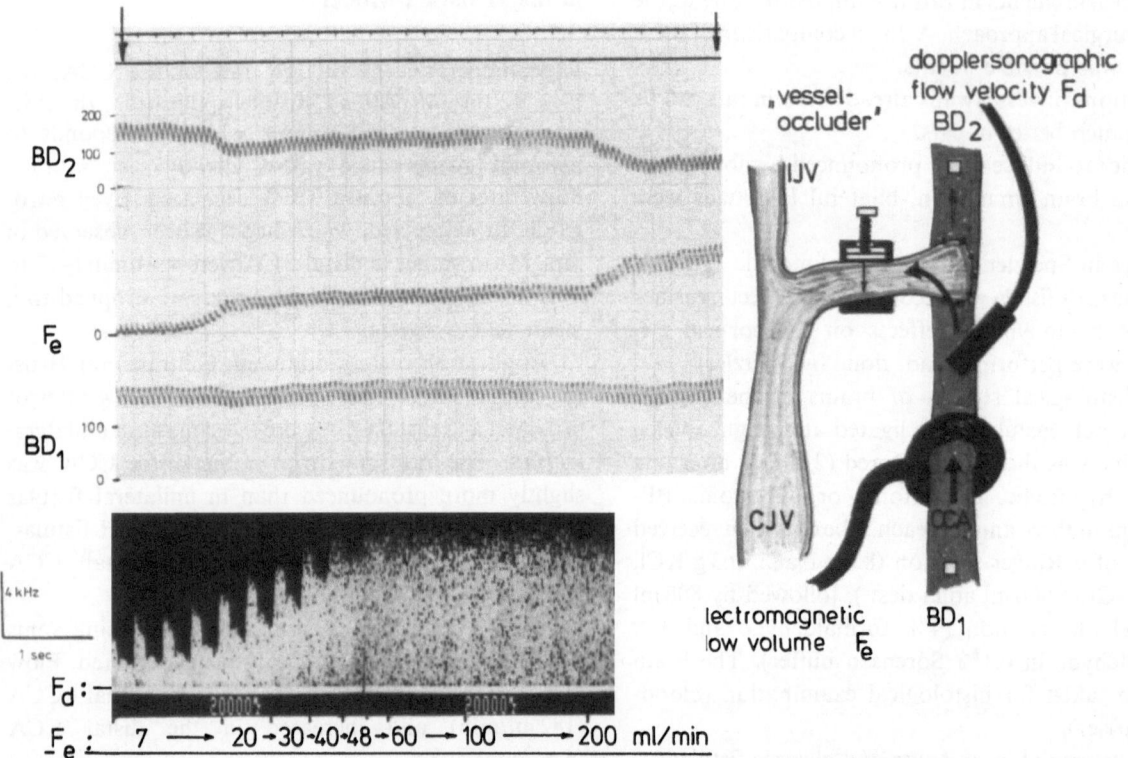

Fig. 28. Comparison of electromagnetic and Doppler recordings in slow and fast flow in the proximal CCA: Progressively opening the H-fistula, the electromagnetic flow rate (*Fe*) and Doppler frequencies (*Fd*) increase. Blood pressure in the distal CCA (*BD 2*) decreases. Note that the upper recording limit of the Doppler device is reached at 48 ml/min of flow rate

4.2. Dopplersonographic Measurements

Flow velocities occurring in the fistulas once again exceeded the recording range of our prototype device.

4.3. Measurements in Varying Fistula Flow

Using a "vessel occluder" (an inflatable fork-shaped cuff around the vessel), we were able to shift flow rates and velocities within a range from completely normal to full fistula flow (Fig. 28). The following results were found in the cats' 1.5 mm thick CCA:

Electromagnetic recordings demonstrated very high, pulsatile flow velocities and flow rates. Flow rates reached 200 ml/min, which mathematically correspond to a flow velocity of 188.6 cm/sec. Such velocities have in fact been found by transcranial Doppler sonography in human angioma feeders. As a characteristic, increases of diastolic flow rates were more pronounced than those of systolic flow.

Such high flow velocities were not detected by our Doppler device. Above values of 48–60 ml/min (45.3–56.6 cm/sec), the displayed Doppler signal turned into nonpulsatile, band-like frequency spectra.

Measurements of orthograde and retrograde flows

Using the vessel occluder, we were able to produce varying flow rates, flow velocities and flow directions in the distal CCA (H-fistula). With stepwise throttling of the proximal CCA, the flow in the distal CCA decreased first to an oscillating, then to a reversed flow (Fig. 29 a, b; Fig. 30).

Results:

a) The Doppler device used cannot unequivocally detect oscillating flows. All flow directions cause an upward deflection on the Angioscan.

b) Using the electromagnetic method, oscillating flows were recorded precisely (Fig. 31).

c) Retrograde flow in the distal CCA showed a nonpulsatile, constant pattern in both recording methods (Fig. 31).

d) With both proximal CCAs occluded, flow velocity spectra consisted of damped systolic 4 kHz waves (Fig. 32). This flow has lost its arterial character. It originates from the vertebral arteries and is attenuated on its way through the Circle of Willis and its many

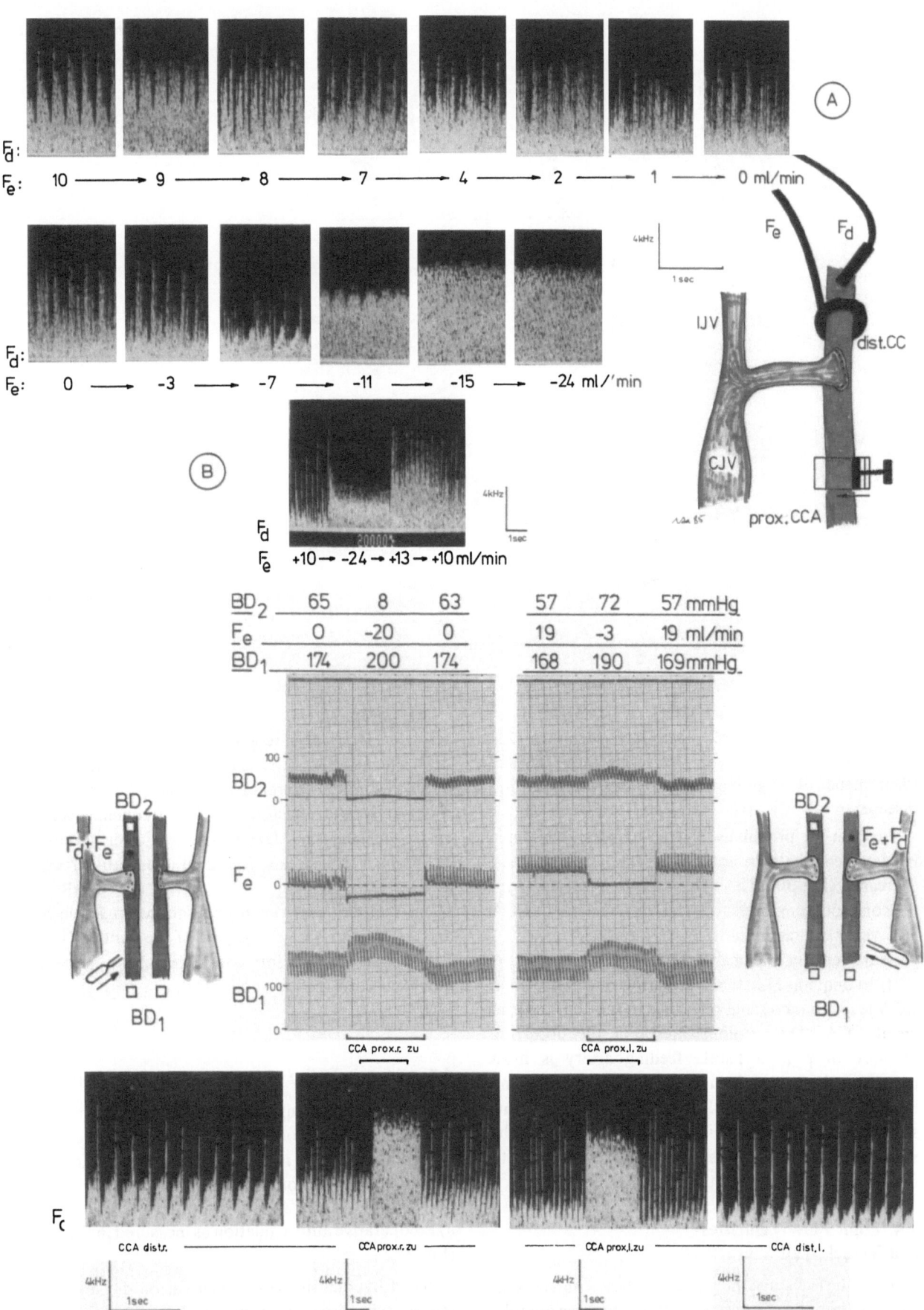

Fig. 29. Comparison of electromagnetic measurements of flow rate (*Fe*) and Doppler recordings of flow velocity (*Fd*) in the distal CCA: A) Progressive throttling of the proximal CCA converts the orthograde pulsatile distal CCA flow into a retrograde and nonpulsatile flow in the Doppler device used (Angioscan), ortho- and retrograd flows are displayed by upward deflections. B) Clipping of one proximal CCA in bilateral H-fistula (*BD 1* pressure in proximal CCA, *BD 2* pressure in distal CCA)

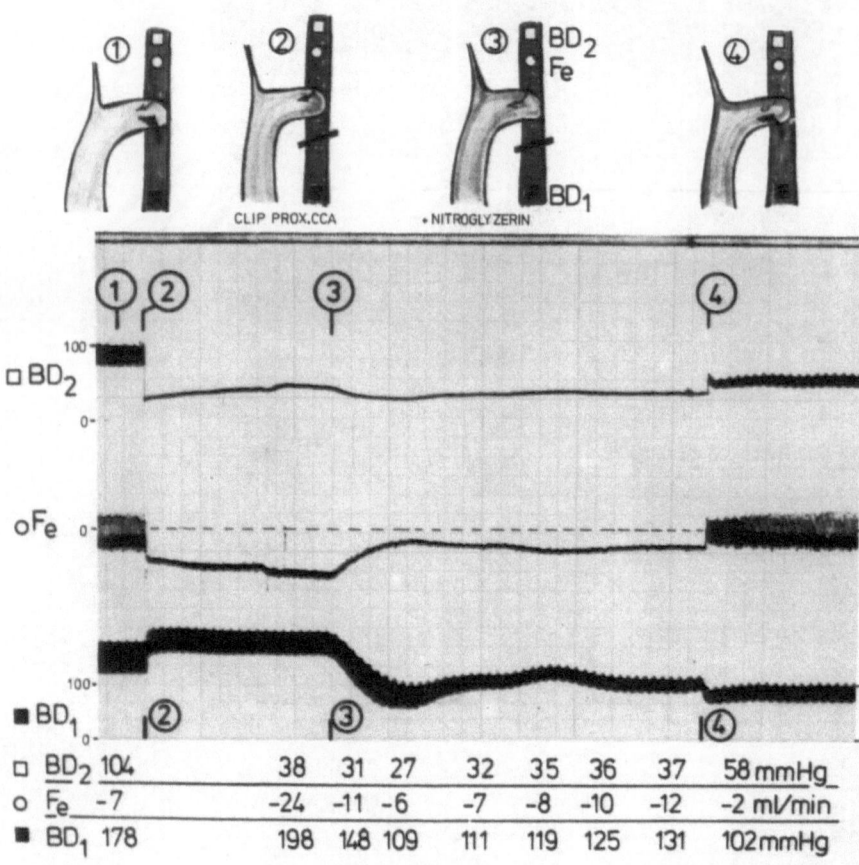

Fig. 30. Continuous electromagnetic recording (*Fe*) under varying conditions: *1* fistula open, *2* proximal CCA occluded, *3* proximal CCA occluded and systemic hypotension, *4* fistula reopened and systemic hypotension (*BD 1* pressure in proximal CCA, *BD 2* pressure in distal CCA)

	②	③						④	
□ BD₂	104	38	31	27	32	35	36	37	58 mmHg
○ Fₑ	-7	-24	-11	-6	-7	-8	-10	-12	-2 ml/min
■ BD₁	178	198	148	109	111	119	125	131	102 mmHg

branches. In transcranial Doppler sonography of human cerebral vessels, such low pressure (20–30 mmHg) and low pulse flows are seen in steal phenomena of angiomas and in occlusive carotid disease.

e) With the proximal CCA occluded and systemic BP lowered, the retrograde flow in the distal CCA decreased (Fig. 30). This can be easily explained with the corresponding pressure drop in contralateral CCA and vertebral circulation. Nevertheless, distal CCA pressure remained unchanged.

f) In bilateral H-fistulas, occlusion of one proximal CCA leads to increasing pressures in the contralateral distal CCA. This demonstrates that the occurring pressure drop in a fistula feeding artery is most pronounced at the fistula itself. Brain supplying branches leaving proximally from the fistula only show small pressure drops. This is considered one of the shortcomings of Spetzler's model when transferred to the human angioma situation (Fig. 29).

4.4. High Flow Velocities and Vessel Wall Vibrations

Close to the H-fistula, the draining external jugular vein shows vibrations of its vessel wall that become audible,

visible and palpable. They result from enlarged vessel diameter, high pressure gradient, thin vessel wall, high flow velocity and high flow rate. Turbulencies may occur at different parts of the fistula (Fig. 33).

Next to the anastomosis, low frequency venous vibrations below 500 Hz are visible. Faster vibrations of about 1,000 Hz are found at the fistula feeding artery.

In long-term experiments, vibrations cause scleroses and stenoses in the venous part of the fistula. The H-fistula model is therefore considered also a model for atherosclerosis (Staubesand 1979).

4.5. Autoregulation in Acute and Chronic H-fistulas

Blood pressure changes were medically induced with Effortil´ or Trinitrosan´. Flow rates were measured electromagnetically in order to investigate the autoregulative capacity of the CCA.

a) Preoperative autoregulation as measured in the CCA (Fig. 35):

Induced hypotension and hypertension did not cause relevant changes of flow rate in the normal CCA.

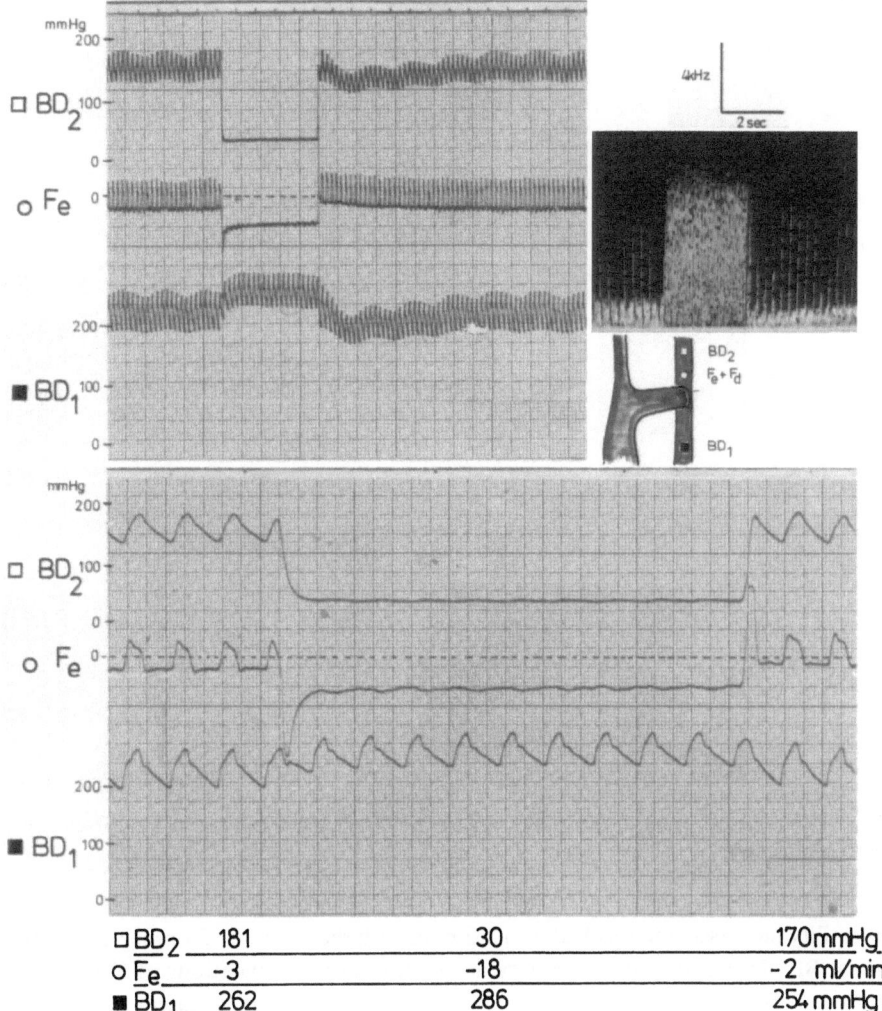

☐ BD₂	181	30	170 mmHg
o Fe	-3	-18	-2 ml/min
■ BD₁	262	286	254 mmHg

Fig. 31. Flow rate (*Fe*) and flow velocity (*Fd*) in distal CCA, pressures in proximal (*BD 1*) and distal (*BD 2*) CCA: With occlusion of the proximal CCA, the systemic blood pressure increases, whereas BD2 drops. The reversed flow in the distal CCA is nonpulsatile. Bottom: slow registration

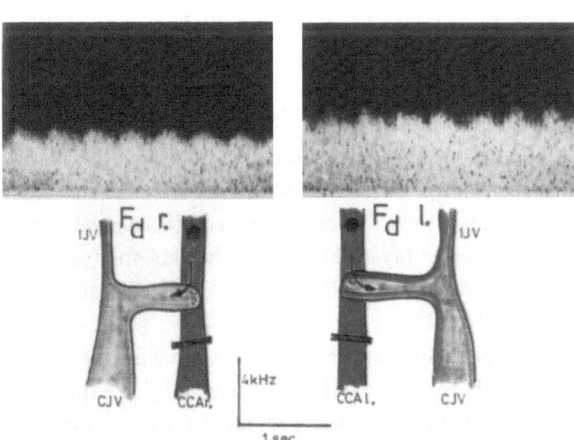

Fig. 32. Doppler recording (*Fd*) of distal CCA in bilateral H-fistula with both proximal CCA occluded: damped systolic waves of retrograde flow

b) Autoregulation as measured in proximal CCA in H-fistulas (Fig. 36).

The flow rates were pressure dependent in this part of the vessel.

c) Autoregulation as measured in distal CCA in H-fistula: (Fig. 37)

A drop in systemic BP caused increasing flow rates in the distal CCA which subsequently returned to previous values with normalization of BP. Pulsatility of the electromagnetically measured flow is most pronounced with the lowest BP. Rising BPs result in decreasing pulsatility while the flow rate remains stable.

d) Autoregulation as measured in distal CCA in bilateral H-fistula:

Hypotension led to an initial slight increase of flow rate in the distal CCA, followed by a significant decrease.

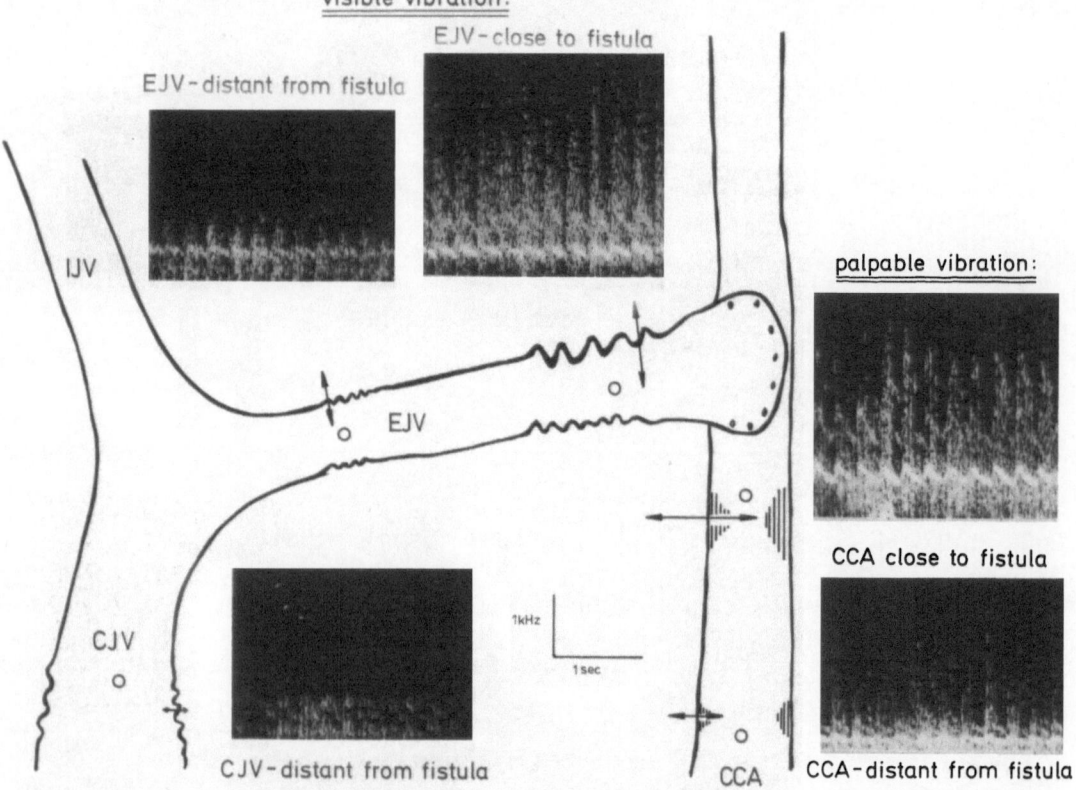

Fig. 33. Doppler recordings of different vessel segments in H-fistula: Vibrations become visible (veins), palpable (arteries) and also audible. They are most pronounced in the close vicinity of the anastomosis

Pulsatility increased and reached its maximum with the lowest BP. Rising BP resulted in decreasing pulsatility and flow rate.

e) Autoregulation as measured in the CCA immediately after occlusion of chronic bilateral fistula (2 months):

Lowering the BP to 65 mmHg did not affect flow rates in the CCA after bilateral fistula occlusion. Pulsatility of the electromagnetic flow increased. Disturbances of autoregulation were not found. Raising the BP did not change CCA flow rates either. Pulsatility was slightly reduced. Autoregulation remained intact (Fig. 38).

These findings suggest undisturbed autoregulation as measured in feeders of a two-month old bilateral fistula. The results are in contrast with Spetzler's theory of abolished autoregulation of brain supplying arteries in cats.

4.6. Measurements of Cortical Perfusion and Flow Rates

In four cats, dopplersonographic measurements of flow velocity in cortical arteries were carried out. In addition, cortical microcirculation was studied with the laser Doppler technique. Thereby, hemodynamic implications of H-fistulas for cortical perfusion could be investigated.

Measurements were performed under the following conditions:

— Fistula patent and occluded.

— Proximal CCA patent and occluded (reconstruction of Spetzler's model) (Fig. 39).

Results

a) Opening and occluding the fistula (Fig. 40)

With the H-fistula patent, systemic BP was 150/100 mmHg. Pressure distal to the fistula was 60/40 mmHg. Flow velocity * in cortical vessels was about 1.5 kHz (systole). Microcirculation (relative measurements) was 42% of the measuring range (10mV = 100%).

After occlusion of the fistula, systemic BP increased by 5%. BP in the distal CCA reached 2.5 times the former value and was 160/115 mmHg. Flow velocities

---

* Flow velocities measured intraoperatively can only be given in kHz.

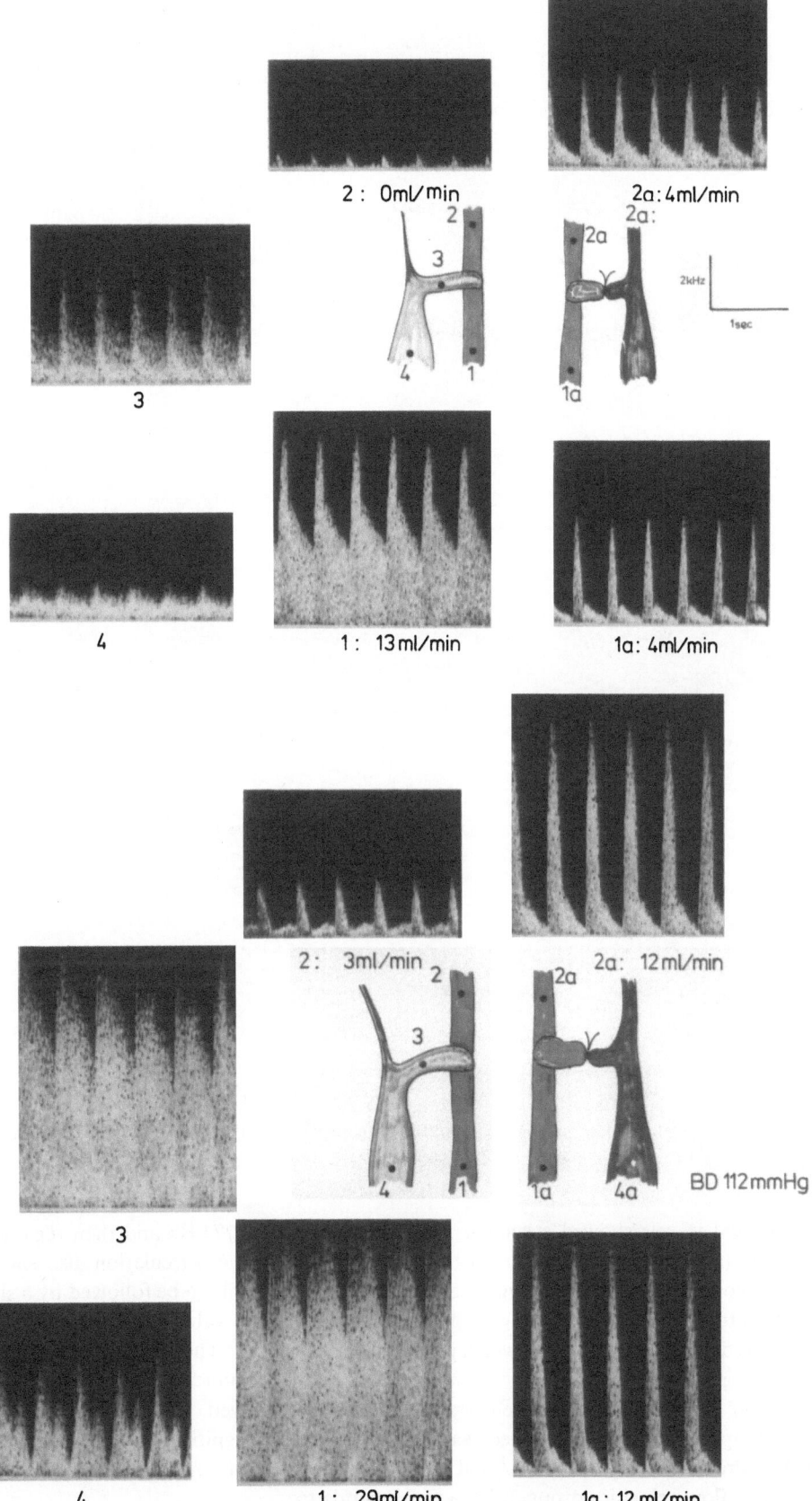

Fig. 34. Autoregulation under varying blood pressure in right-sided H-fistula: Hypertension results in systolic and diastolic flow increase in right CCA; the contralateral CCA shows a dissociated response with increasing systolic and decreasing enddiastolic velocities

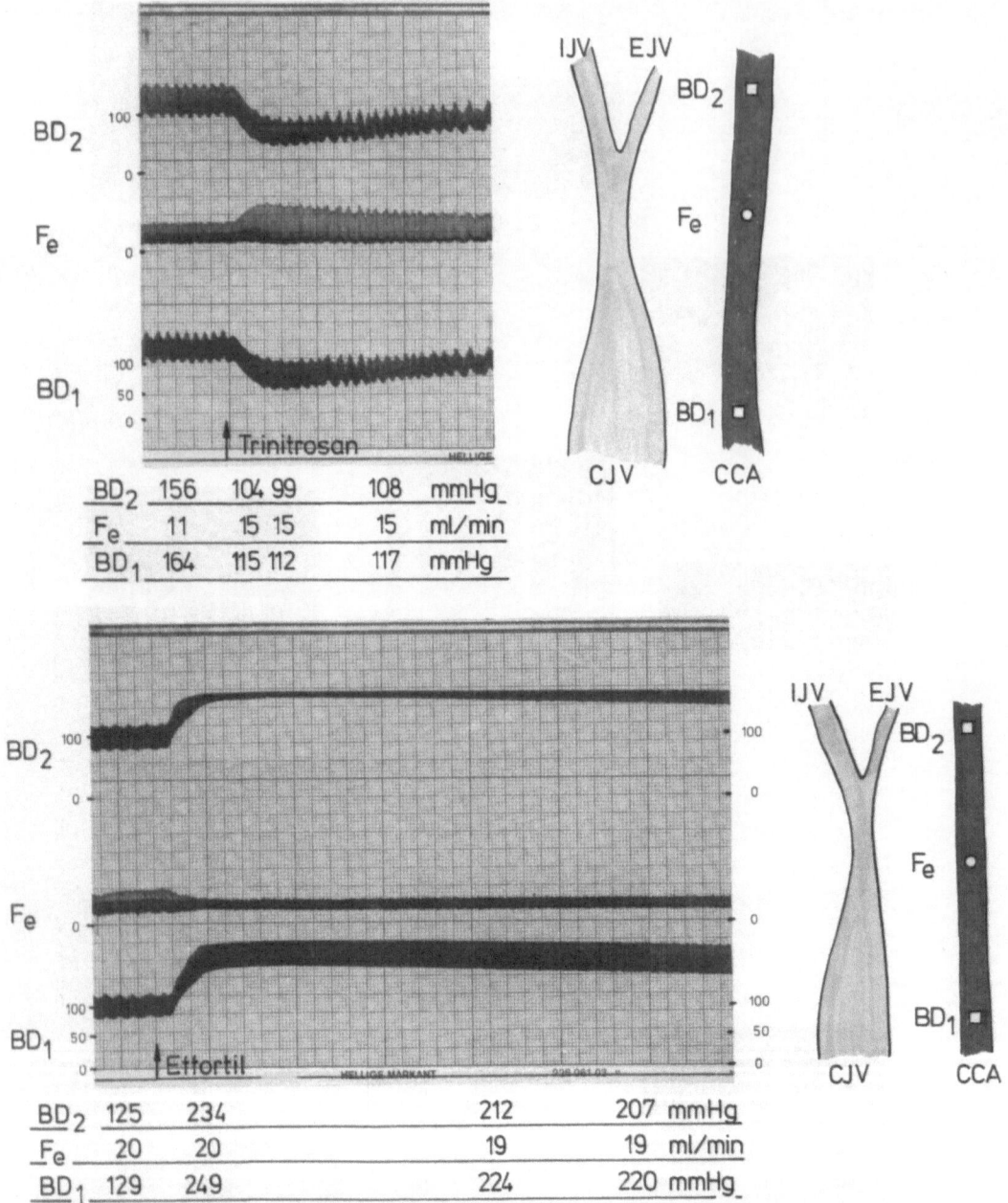

Fig. 35. Autoregulation under varying blood pressure in normal CCA: Hypotension causes a slight increase in flow rate (*Fe*), whereas an additional slight reduction is induced by hypertension (*BD 1* systemic blood pressure, *BD 2* distal CCA pressure)

in cortical vessels showed a sudden 53% increase to 2.8 kHz (systole) and then returned to normal within five seconds. Cortical microcirculation initially increased to 80% of the measuring range and within ten seconds dropped back to 56%, which is 14% above the value in patent fistulas.

After a steady state had been reached, the fistula was reopened. Systemic BP decreased to its initial value. Pressure in the distal CCA fell to 50/40 mmHg. Cortical flow velocities dropped for a few seconds to

0.77 kHz and then regained previous values. Cortical microcirculation also showed a sudden drop to 28% only to be followed by a short peak and then its return to values found in patent fistulas (42%).

These experiments suggest that significant hemodynamic effects of unilateral H-fistulas are only short-lived. Cortical microcirculation is reduced by 25%. Significant impairment of cortical perfusion, however, does not occur.

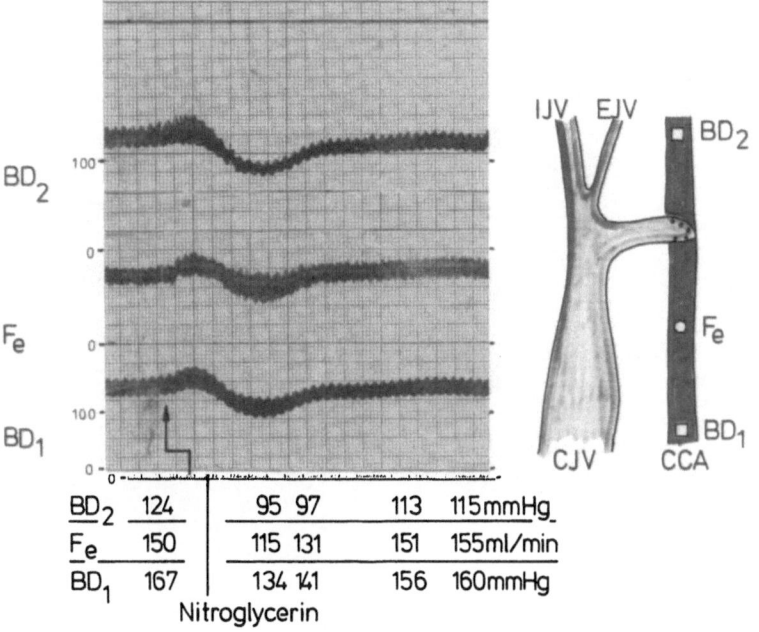

BD$_2$	124	95	97	113	115 mmHg
Fe	150	115	131	151	155 ml/min
BD$_1$	167	134	141	156	160 mmHg

Nitroglycerin

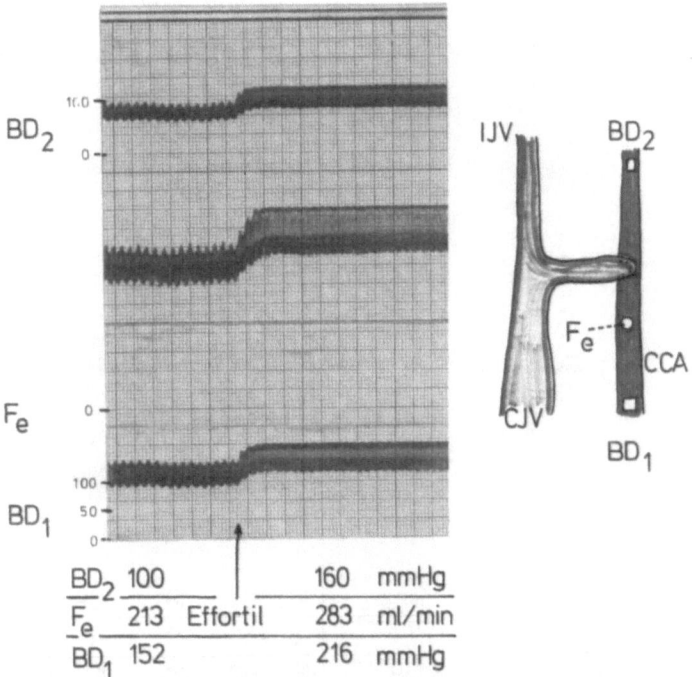

Fig. 36. Impaired autoregulation as measured in proximal CCA in H-fistula: Flow rates in the fistula supplying vessel (*Fe*) change depending on its blood pressure

BD$_2$	100		160	mmHg
Fe	213	Effortil	283	ml/min
BD$_1$	152		216	mmHg

b) Occluding and opening the proximal CCA (Figs. 41 and 42)

With the fistula patent, systemic BP was 160/100. Pressure in the distal CCA was 80/60. Flow velocity in cortical vessels was 1.4 kHz and cortical microcirculation was 44%.

With occlusion of the proximal CCA, reversed blood flow in the distal CCA was induced with the blood now running into the jugular vein.

Systemic BP then increased by 10% and pressure in the distal CCA dropped to 30 mmHg. Flow velocities in cortical vessels decreased from 1.4 kHz to 1.2 kHz. Cortical microcirculation remained unchanged.

After a steady state had been reached, the proximal CCA was reopened. Systemic BP and distal CCA pressure returned to former values. Flow velocities also regained the previous 1.4 kHz within seconds. Cortical microcirculation was slightly depressed for a period of five seconds.

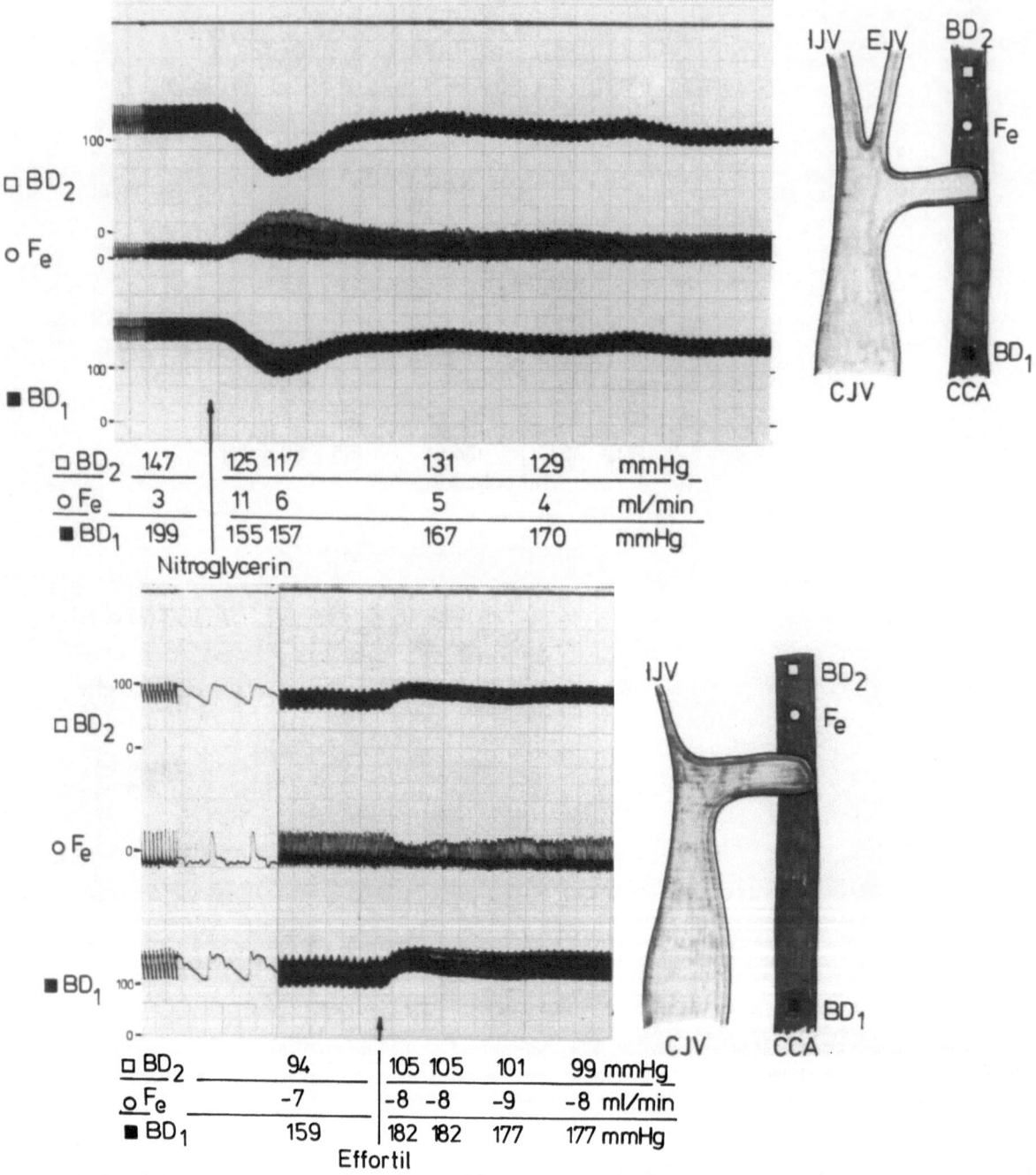

□ BD₂	147	125 117	131	129	mmHg
○ Fₑ	3	11 6	5	4	ml/min
■ BD₁	199	155 157	167	170	mmHg

Nitroglycerin

□ BD₂	94	105 105	101	99 mmHg
○ Fₑ	−7	−8 −8	−9	−8 ml/min
■ BD₁	159	182 182	177	177 mmHg

Effortil

Fig. 37. Intact autoregulation of distal CCA in H-fistula: Flow rates in this vessel, which is connected in parallel to the fistula (*Fe*), show normal regulation under varying blood pressure

These results suggest that Spetzler's model does have a slight effect on cortical flow velocities, but does not disturb cortical microcirculation whatsoever. It therefore seems to be unsuitable for investigations on "normal perfusion pressure breakthrough" phenomena.

4.7. Blood-brain Barrier

Nine animals with bilateral T-fistulas had their fistulas occluded. Three animals each were then exposed to medically induced hypotension and hypertension for one hour. Three animals were left normotonic. As

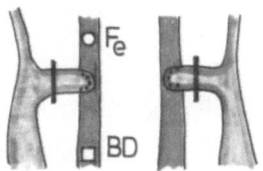

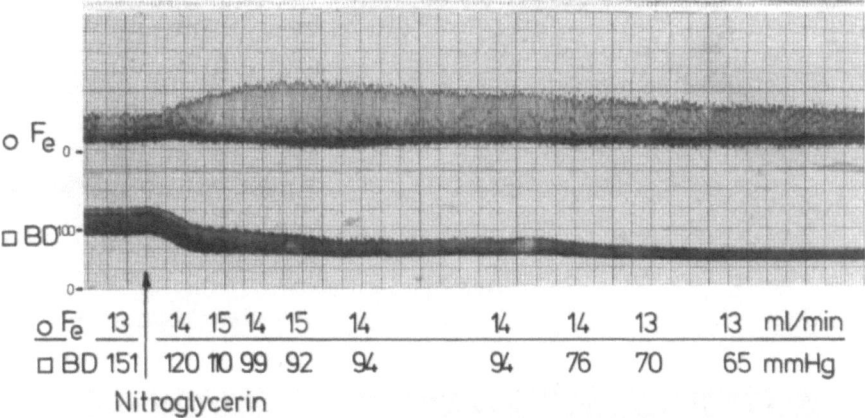

$_{\circ}$ F_e	13	14	15	14	15	14	14	14	13	13 ml/min
$_{\square}$ BD	151	120	110	99	92	94	94	76	70	65 mmHg

Nitroglycerin

Fig. 38. Intact autoregulation after sudden occlusion of an 8-week-old bilateral H-fistula: No relevant changes of flow rate occur in the distal CCA which had been exposed to low pressure conditions for 8 weeks

F_e	12	12	11	10	11	12 ml/min
BD	87	91	100	110	125	130 mmHg

Effortil

Fig. 39. Reconstruction of Spetzler's model: After H-fistula application, the cats were trephined precentrally (1 cm). Following the opening of the dura, the Doppler or laser doppler probes could be attached to the cortex using a micromanipulator (David Instruments No. 1760/S). Occlusion of the proximal CCA then brings about the same conditions as in Spetzler's model

described above, Evans blue was then given to study disturbances of the blood-brain barrier.

Results

— No hemorrhages were found in hormotension and hypotension.

— No cerebral atrophy was found.

— Small extravasations of Evans blue were seen only in hypertensive animals and occurred in all brain regions (in fistula carrying cats as well as in normal animals).

5. Discussion and Summary of Animal Experiments

a) Animal models

The H-fistula model fulfills the following requirements for experimental studies on angioma-like conditions:

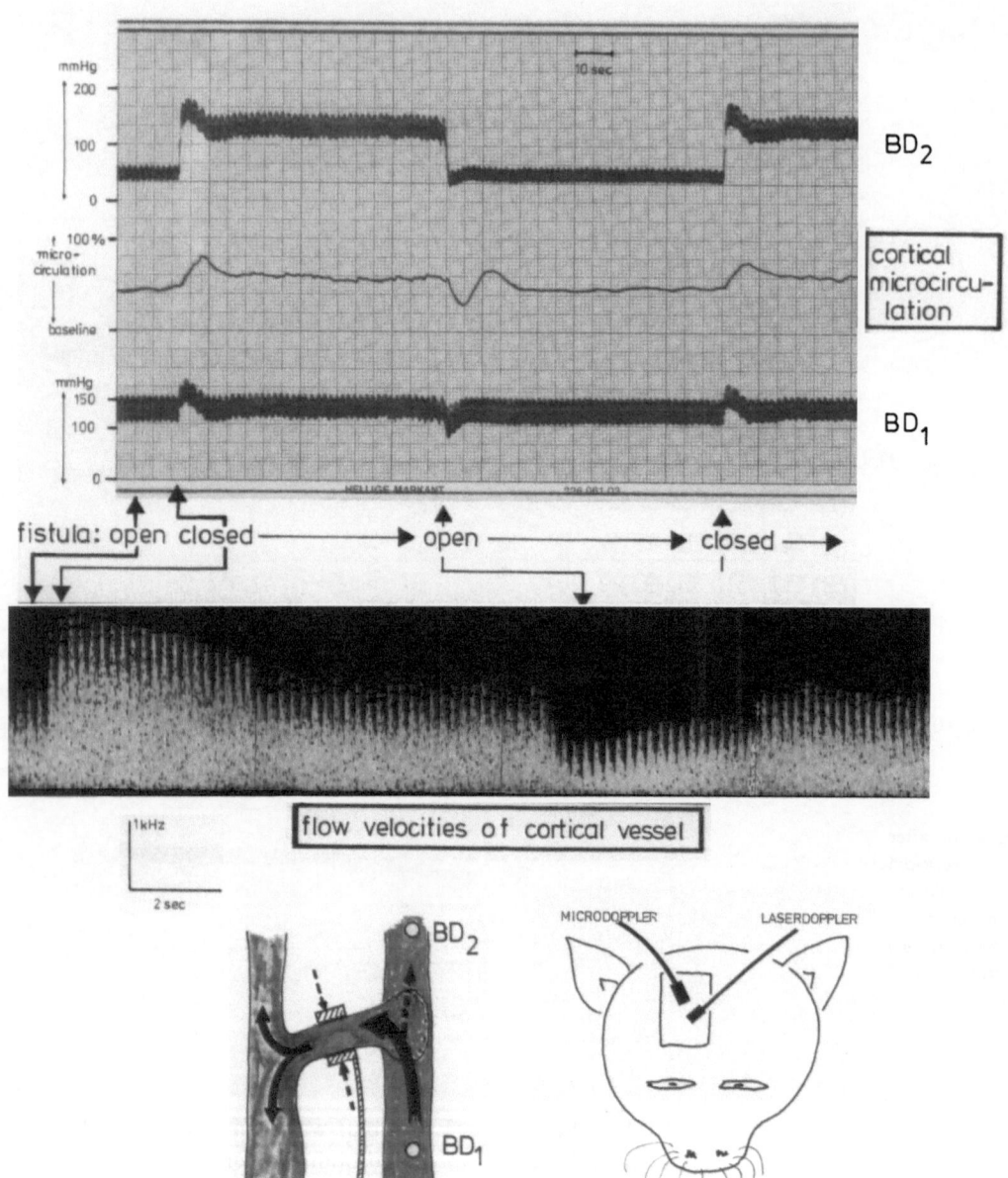

Fig. 40. Recordings of cortical flow velocity (Doppler) and microcirculation (laser Doppler) with the H-fistula open and occluded. The cat's head is fixed, the probes are held by a micromanipulator. The vessel occluder lies in front of the cat

—High shunt flow in the fistula-supplying proximal CCA.

—Low flow rate and low pressure in the brain supplying distal CCA (Fig. 43).

—Venous congestion even in cerebral sinus.

The H-fistula reduces the brain's arterial inflow pressure by 50% on one or both sides compared to only about 10% in Spetzler's model. Another advantage of the H-fistula is that, with progressive throttling of certain vessels, varying flow and pressure conditions can be induced while the systemic BP remains unchanged.

The T-fistula model was predominantly used in our study for investigations of vessel wall changes in response to high blood flow velocity.

b) Flow rates

Following the application of cervical H- or T-fistulas, the electromagnetic proximal CCA flow rates increase to about ten times the initial values in cats and in rats. This corresponds to findings in angioma patients, where angioma feeding vessels also have about ten-fold flow rates compared to normal controls. In the long-

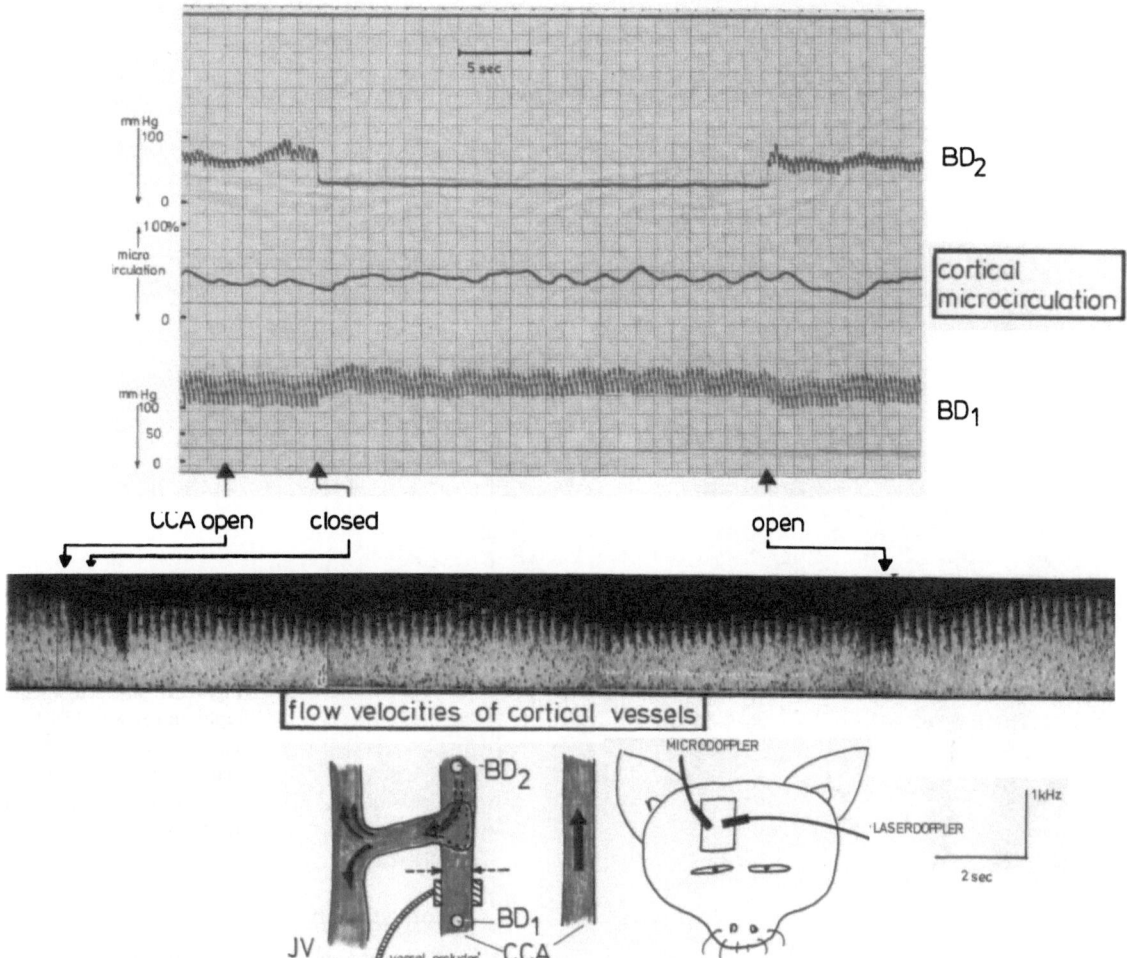

Fig. 41. CCA-occlusion proximal to the H-fistula produces Spetzler's model. Cortical flow velocity (Doppler), microcirculation (laser Doppler) and blood pressure are measured (see text)

term animal fistula, the flow rates tend to decrease as a result of vascular stenoses.

c) Flow velocities

The only Doppler device that was commercially available for intraoperative use (MF 20, EME) allowed measurements of flow velocity up to 84 cm/sec. As velocities in AV fistulas are much higher, only flow direction and patency of the shunt were demonstrable with this technique.

d) Stream resistance

Peripheral stream resistance of the CCA drops after application of cervical AV fistulae. The resistance indices were calculated according to Pourcelot from our electromagnetic flow profiles. As expected, stream resistance was lower in direct (T) fistulae than in the H-model.

e) Vascular changes

Arterial and venous vessel diameters increase after cervical fistula application. Enlargement was much more pronounced in draining veins (up to 250%) than in the feeding arteries (average 23%). Vascular widening under high blood flow conditions is a well-known phenomenon. It is most probably due to low frequency transverse vibrations that reportedly result in an altered composition of elastic and collagenous fibers in the vessel wall (Boughner 1970, Foreman 1970). After fistula application in the rats, the draining veins immediately swelled up and further enlarged during the next two months. Staubesand (1980) studied the occurring vessel wall alterations under the electron microscope. He found changes similar to those in varicosis (Staubesand 1978). The intercellular spaces in the tunica media widen, fibrous elements proliferate, dysplastic collagen, matrix vesicles and muscle cells of the m-type occur.

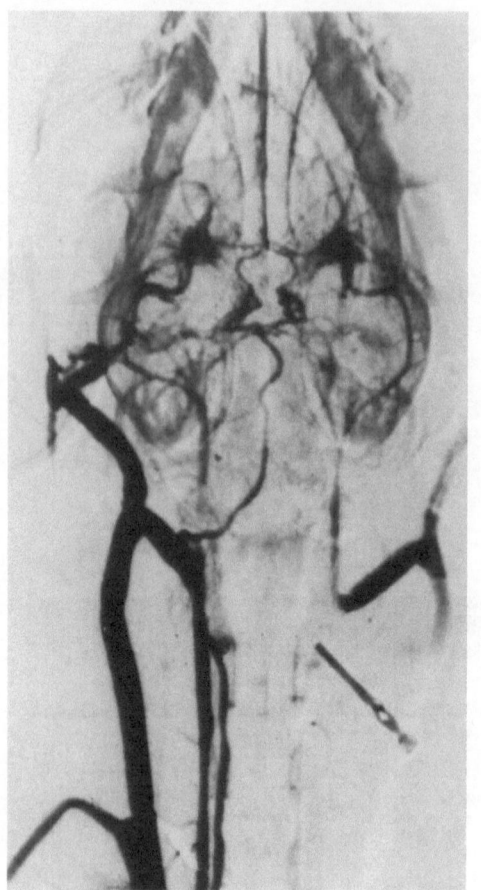

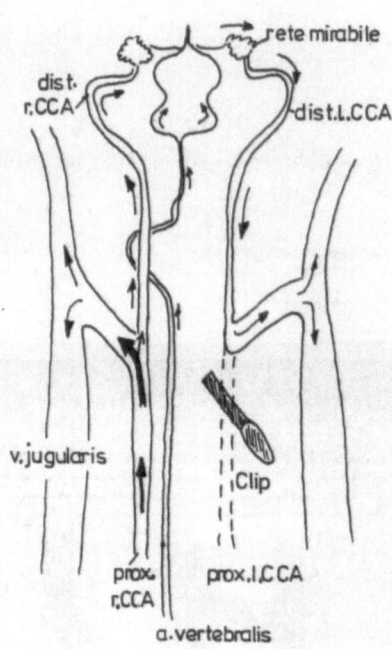

Fig. 42. Right-sided brachial arteriography in cat with bilateral H-fistula: The proximal left CCA is clipped, so that the flow from right CCA runs into the right H-fistula and also intracranially to the Circle of Willis, from where the distal left CCA is filled

f) Autoregulation

Vascular autoregulation of the brain remains intact even after long-term cervical AV fistula application. CCA flow rates in our 15 cases did not change with induced hypotension or hypertension after the cervical fistula had been suddenly occluded. This is in contrast to Spetzler's experimental data of impaired autoregulation in 5 out of 25 cats after sudden occlusion of their cervical shunts.

g) Cortical perfusion

Cortical perfusion as measured by the laser Doppler method is reduced by 25% in the cervical H-fistula model only. Spetzler's model was tested by us and found to have only very slight effects on cortical perfusion.

h) Blood brain barrier

Alterations of the blood brain barrier (Evans blue method) in chronic cervical AV fistulas are only seen when the systemic BP is elevated above the limit of vascular autoregulation. Such changes then correspond to those found in hypertensive encephalopathy (Auer 1978).

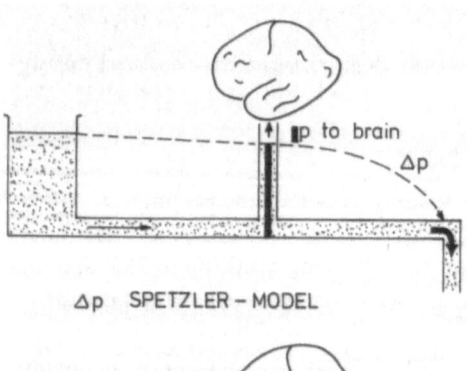

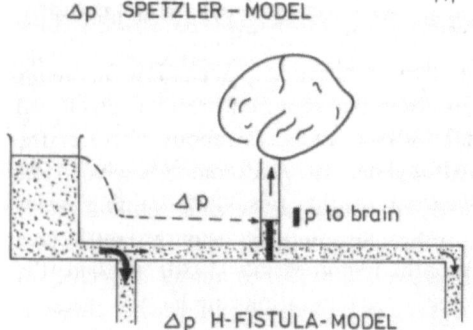

Fig. 43. Nornes' construction applied to the conditions of Spetzler's/Scott's model and our H-fistula. The pressure gradient △ p of brain supplying branches is much lower in the H-fistula model because the major pressure drop occurs proximal to these vessels

III. Angiography in Angioma Patients Before and After Surgery

Cerebral intravascular pressure rises after angioma exclusion. This is said to cause global and local hyperperfusion of the brain that may threaten the patient's outcome. (Mullan 1973).

Preoperative and postoperative cerebral angiograms were evaluated in reference to signs of such hyperperfusion or hyperemia. We thereby tried to gain predictive criteria for hemodynamically critical cases which could later present intraoperative or postoperative complications.

Postoperative angiography was performed immediately after surgery with the patient still in general anesthesia. We originally thought that remaining parts of the angioma were responsible for postoperative hemorrhage. However, the one case that underwent incomplete removal did not bleed, whereas other patients did. We therefore tried to establish other criteria for predicting hemorrhages.

1. Materials and Methods

Thirty-five patients underwent immediate postoperative angiography (Table 1). Three of them were investigated again after 8–15 days. The approximate angioma volume was determined from the sagittal and frontal views. It ranged from 1–80 ccm.

Besides number, length and diameter of the feeding arteries and draining veins, preoperative blood distribution was evaluated. We developed the following classification:

1. Visualization of angioma; normal filling of cerebral arteries.
2. Visualization of angioma; reduced filling of cerebral arteries.
3. Visualization of angioma; very scarce or no filling of cerebral arteries.

The following parameters were evaluated in postoperative angiograms: number, diameter and length as well as duration of contrast filling of so-called "stagnating arteries". The term "stagnating artery" was applied to former angioma feeders that show very slow flow

Table 1. *Localization of the angiomas (n = 35)*

Frontal	4
Fronto-parietal	6
Parieto-occipital	9
Temporal	3
Sylvian	4
Midline	5
Ventricle	4

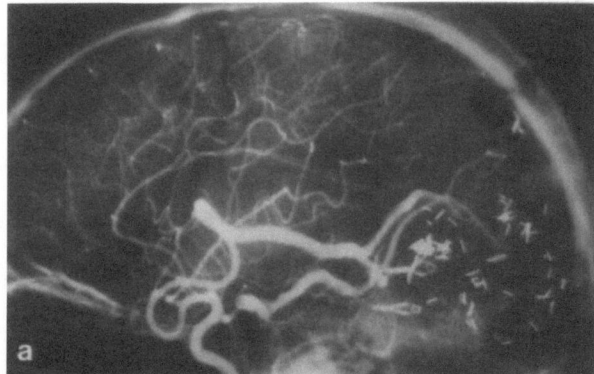
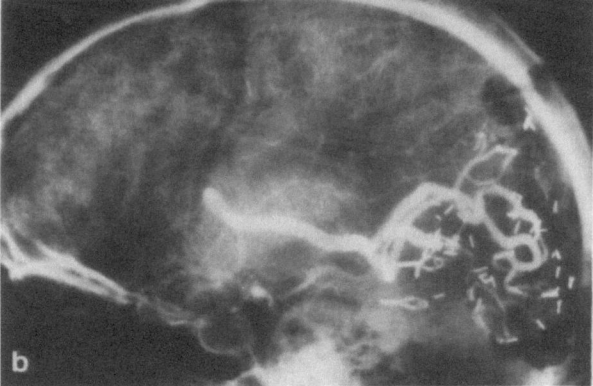

Fig. 44. Stagnating vessels after removal of an occipital angioma: a) arterial phase, b) capillary phase: branches of MCA and PCA are still enlarged; they show very slow drainage and remain visible over 10 sec

and remain visible even in the capillary and venous phase of the angiogram.

In addition, the diameter of the internal carotid artery was determined in its extradural portion at the

Table 2. *Angiographic Changes of Hemodynamics After Excision of Angioma (n = 35)*

Filling of cerebral arteries on carotid angiography	Number of cases	
ACA + MCA ipsilateral	8	
ACA + MCA ipsilateral + contralateral ACA	7	
ACA + MCA ipsilateral + contralateral ACA + MCA	2	
Panangiography	10	
No changes*	6	

* Very small angiomas (4). Angiomas of corpus callosum (2).

siphon. Thereby, diameter changes far away from the surgical site should be detected.

2. Results

a) 34/35 patients showed complete AVM exclusion. One small angioma had been partially removed.

b) After AVM-removal, hemodynamics changed considerably in 27 patients (Table 2). Eight patients showed normal filling of ipsilateral arteries that had been previously poorly visualized. The contralateral ACA was visible in seven patients, and the contralateral MCA in two patients. Ten patients showed filling of all cerebral vessels.

c) No significant change was encountered in eight cases. They all had either small or callosal AVMs.

d) The carotid siphon diameter grew smaller after AVM exclusion in 33/35 patients (Table 3).

This reduction correlated with the previous angioma volume. In volumes above 30 ccm, diameters decreased by 24–40% (10 patients). With volumes from 1–30 ccm, reduction was up to 16% (15 patients). No patient showed increasing extradural ICA diameter.

e) "Stagnating arteries" were found in all early postoperative angiograms when the angioma volume had exceeded 4 ccm. These vessels represent former angioma feeders, show large diameters and slow flow after the malformation has been removed. The diameter of stagnating arteries is the same as it was preoperatively. Their length in parietooccipital AVMs often reached 10 cm; angiographic visualization persisted for more than 10 seconds (Figs. 44, 45).

Late postoperative angiography in 1 out of 3 cases

Table 3. *Correlation Between Seize of Angioma and Diameter of ICA*

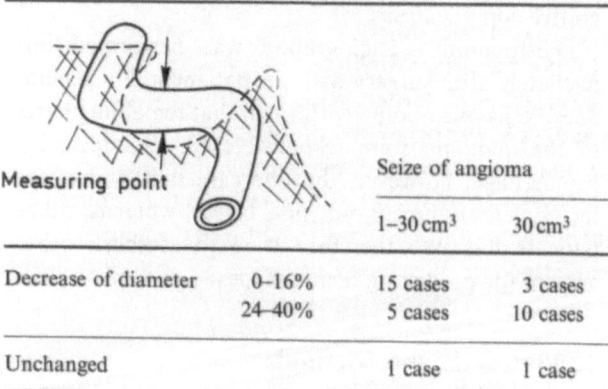

		Seize of angioma	
		1–30 cm³	30 cm³
Decrease of diameter	0–16%	15 cases	3 cases
	24–40%	5 cases	10 cases
Unchanged		1 case	1 case

revealed stagnating arteries, even twelve days after surgery. This patient had a hemorrhage at that time.

f) Complications occurred in ' 12/35 patients (Table 4). In five cases, hemostasis became more and more impaired with advancing AVM exclusion. Nine patients developed brain swelling, which in five cases was local and in four extended over the whole hemisphere. One patient suffered from a sudden, massive intracerebral bleeding that occurred intraoperatively and in close vicinity of the angioma. Early postoperative bleeding at the surgical site was demonstrable in four patients. Late bleeding occurred in two cases on the twelfth and thirteenth postoperative day, respectively. Both underwent reangiography after that event.

g) There is a correlation between postoperative angiographic findings and impending complications. They occur in patients with stagnating arteries of more than 5 cm length and more than 10 seconds of angiographic filling. Complications also ensued in patients

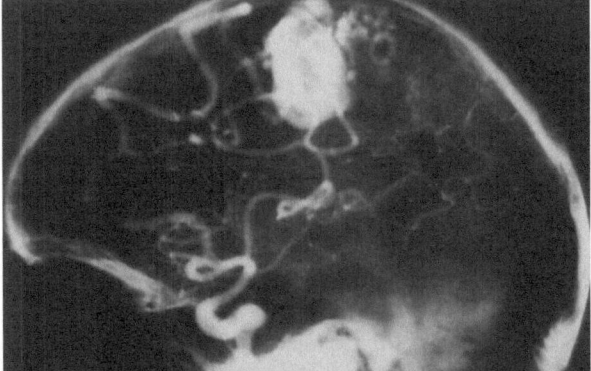

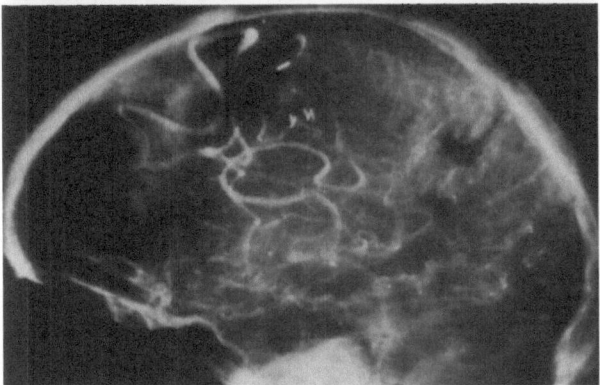

Fig. 45. Pre- and postoperative angiogram in a patient with a left central angioma. Previous feeders become "stagnating arteries" following the AVM removal

Table 4. *Problems During and After Excision of the Angioma in 22 of 35 Patients*

		Number of cases
Intraoperative:	brain edema	3
	intracerebral hemorrhage	1
	problems with hemostasis	5
Postoperative:	rebleeding early (day of excision)	4
	late (after 12 and 13 days)	2
	brain edema local	6
	hemisphere	4

with enlarged pial arteries in the vicinity of the former AVM and in cases with preoperative angiographic steal phenomena (Tables 5 and 6) (Fig. 46).

h) Hyperperfusion or local hyperemia was not observed.

3. Discussion

Exclusion of an angioma results in hemodynamic changes both in its immediate vicinity and the whole brain. Angiographic steal phenomena disappear (Feindel 1975) (Table 6). This may be a sign of im-proved perfusion in previously undersupplied brain regions (Berger 1975, Folkow 1971, Murphy 1954 and Nornes 1980).

Adaptation of vessels in the surroundings of the former AVM is often said to be impaired and delayed. The so-called "stagnating arteries" exhibit slow flow and high pressure, which could be confirmed by intraoperative measurements (see Chap. V). Pressure induced rupture of small vessels may be a cause of postoperative bleeding. Hyperperfusion (Folkow 1971) or breakthrough phenomena (Nornes 1980, Spetzler 1978) in the vicinity of the former AVM, however, could not be demonstrated by means of angiography. We therefore feel that postoperative complications are due to high intravascular pressure in thin-walled, surgically coagulated vessels around the former angioma nidus. As stagnating arteries persist up to

Table 5. *Angiomas with Hemodynamic Risks*

Doppler:	large vessels with low flow and slow diastolic flow (flow resistance ↑)
	high flow angioma
	external type in normal brain vessels
Angiography:	stagnating arteries: visible longer than 10 s length more than 6 cm
	enlarged pial vessels around area of excision
	preoperative steal

Table 6. *Results of Immediate Postoperative Angiography Following AVM Removal*

Former angioma vessels:	1. unchanged diameters of former angioma feeders 2. "stagnating arteries" (large diameter and slow flow) 3. disappearance of large draining veins
Brain supplying vessels:	1. disappearance of steal effects, normal blood distribution 2. visualization of contralateral vessels 3. diameter reduction of ICA siphon 4. no signs of hyperemia

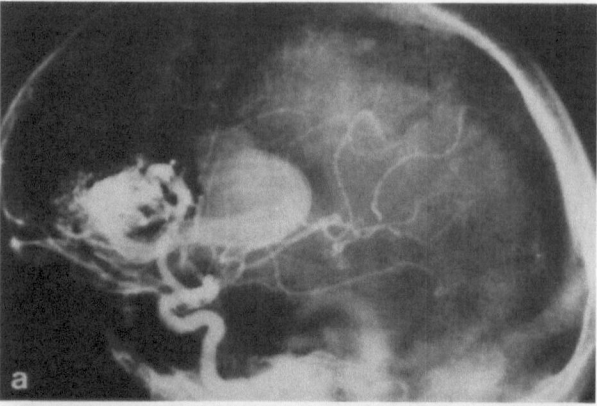
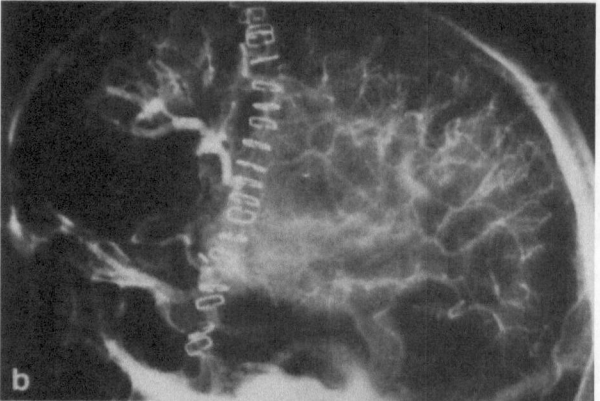
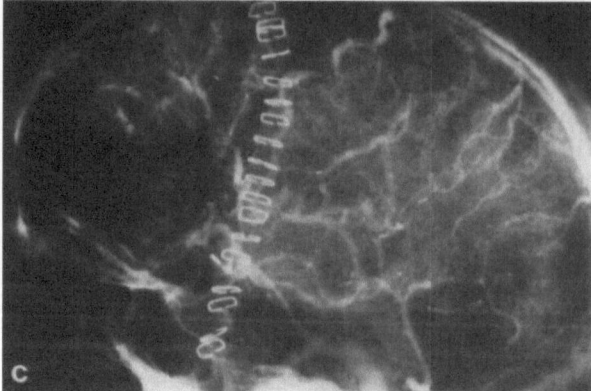

three weeks (Norlen 1949), the risk of bleeding continues throughout this period.

4. Summary and Conclusion

Local and global cerebral hemodynamics change after removal of an angioma (Table 6). The diameter of basal arteries decreases, indicating a change in blood distribution. The diameters of former AVM feeders nevertheless remain relatively stable in the early postoperative phase. Adjustment to normal conditions takes up to three weeks. Critical angiomas can be identified angiographically. If hemodynamic complications occur, they are due to increased postoperative intravasal pressure (Fig. 47). Hyperperfusion is not found in postoperative angiograms.

Critical AVMs should be treated in special ways. Postoperative controlled hypotension, step-wise removal or preoperative partial embolization have to be considered. Feeders should be clipped or ligated but not only coagulated.

Fig. 46. a) Left frontolateral angioma with aneurysmatic enlargement of draining vein. b) and c) Postoperatively enhanced visualization of pial arteries in the former AVM's surroundings

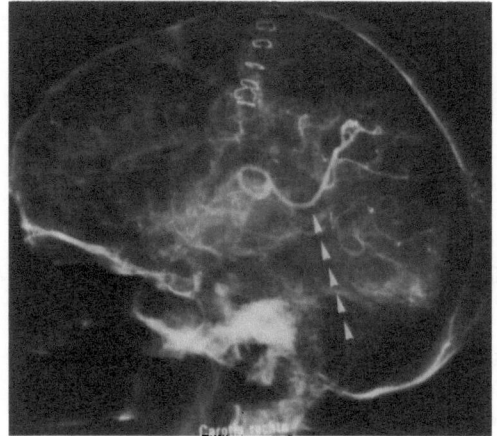

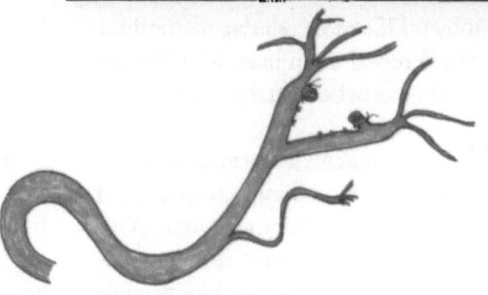

"stagnating artery"

high pressure   large diameter   low flow

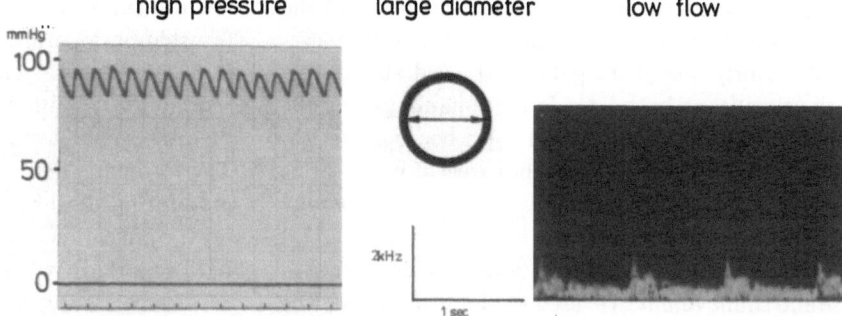

Fig. 47. Angiography after removal of a small angioma showing "hyperemia" in the operation site. Below: characteristics of "stagnating arteries": large diameter, high pressure, low flow

IV. Transcranial Doppler Sonography in Angioma Patients

1. Introduction, Methods and Normal Values

1.1. Transcutaneous Recording Methods

Angiography is still by far the most reliable method for the exact diagnosis of cerebral angiomas. Many other transcutane procedures have been used, but are of lesser significance.

Arteriovenous O_2-difference was measured to obtain information on the hemodynamic relevance of cerebral AVM (Lassen 1964, Shapiro 1966). This method is no longer used today. The first dopplersonographical studies were made in the mid-70's when extracranial blood flow velocities were recorded in angioma patients (Büdingen 1978, Nies 1976). In only 37.5% of these patients was a correct diagnosis of the side of the AVM possible with extracranial measurements. Thirty-one of the patients showed abnormally high velocities on both sides. In the remaining cases, the findings were normal on both sides because cervical vessels may have enlarged and the Circle of Willis forms the pool from which blood is distributed peripherally (von Kalckreuth 1985).

Ancri (1985) tried to classify angiomas according to size and shunt volume. He used a pulsed Doppler device and measured flow velocities of the internal carotid arteries. The resistance indices were low and diastolic flow velocities elevated. Ancri introduced a special index ("diastolic fraction") which, however, did not prove its worth.

Transcranial CBF-measurements allow the rough estimation of regional alterations of blood distribution. As the local resolution of these methods is poor, areas of single brain arteries cannot be reliably identified. Somewhat better results can be obtained by the dynamic CT-method (Axel 1980, Coin 1977, Dobben 1979, Dunn 1979). With the electronic stethoscope, acoustic signals became detectable (Olinger 1977, Sekhar 1984). With transorbital auscultation, aneurysms show characteristic frequencies ("spikes"), whereas angiomas were found to give broad "bruits" in about 60% of the cases. A predecessor of today's transcranial Doppler was the pulse echo method (Leksell 1956, Freund 1965, Kristensen 1967). For the first time, transcranial recordings of sound emissions from vascular pulsations became possible, but the vessels could not be identified.

In the 1980's, the first transcranial Doppler recordings were made in infants with open fontanels. For the first time, intracranial flow velocities became measurable, which proved to be useful especially in galenic angiomas (Strassburg 1982, Mullart 1982, Sivakoff 1982).

In 1982, a new Doppler method was reported. Using low transmission frequencies and pulsed performance, transtemporal, transorbital and transnuchal evaluations of all basal cerebral arteries became possible (Aaslid 1982). With this device noninvasive assessments of intracranial hemodynamics have been facilitated. The technique now provides the method of choice in monitoring cerebral vasospasm (Aaslid 1984). Until now, applications in angioma patients have had not been reported.

There is only one report in the literature on the intracranial hemodynamics in human cerebral angiomas (Nornes 1980). It is based on intraoperative recordings of intravascular pressure and blood flow velocities in angioma supplying arteries. Especially in AVM patients, however, preoperative information on impaired hemodynamics is desirable. In our study, we attempted to ascertain whether an unequivocal diagnosis of cerebral AVM is possible using the transcranial Doppler. We also tried to evaluate the hemodynamic situation in such patients and devoted special attention to alterations of blood distribution and steal effects from normal brain. Such phenomena are well known from angiography, but represent mere morphological findings of unknown significance for the development of neurological or psychic symptoms.

Cerebral vascular autoregulation in angioma patients was another topic of our investigation. It is not

Table 7. *Methods Used in the Diagnosis of Cerebral Angiomas (Other than Angiography)*

Recording site	Method	Author
Heart	ultrasound	Sivakoff 1982
Cervical vessels	O$_2$-difference ultrasound, Doppler	Lassen 1964 Nies 1976, Büdingen 1978 Kalkreuth 1985, Ancri 1985
Head (transcranially)	measurements of cerebral blood flow	
	NO$_2$ method	Kety and Schmidt 1946, Bernsmeier 1952
	radioactive clearance methods	Lassen 1956 (krypton 85) Häggendahl 1965 (xenon 133 i.a.) Fujishima 1967 (RISA i.v.) Oeconomos 1969 (Tc 93 i.a.) Prosenz 1971 (Tc 99 + xenon 133) Menon 1979 (xenon 133 inhalation) Takeyama 1980 (radioalbumin i.a.) Yamada 1982 (xenon 133 i.v.) Okabe 1983 (xenon inhalation)
	CT and dynamic CT	Hacker 1977, Heinz 1979 Traube 1980, Dobben 1979 Wing 1980, Wasenko 1985
	electronic stethoscope	Olinger 1977, Kosugi 1983 Sekhar 1984
	ultrasound	Leksell 1956, Freund 1965 Kristensen 1967

	ultrasound, Doppler
	   through fontanel
	   through skullbone
		Straßburg 1982, Sivakoff 1982
		Mullart 1982
		Aaslid 1982

known whether AVM themselves have a regulative capacity and to what extent autoregulation in the surrounding brain vessels is impaired. The postoperative course was also of interest. We tried to determine the duration of impaired hemodynamics and reconsidered the general conception of brain hyperperfusion after AVM surgery.

1.2. Transcranial Doppler Sonography

In adults, the skull builds a barrier for the penetration of the usual Doppler frequencies of 5–10 MHz. With emitted frequencies between 1.5 and 3 MHz, ultrasound absorption is lower, so that basal cerebral arteries can be evaluated through thin parts of the temporal squama and the orbit. These "cranial windows" (Aaslid) allow recordings in 95–97% of all subjects.

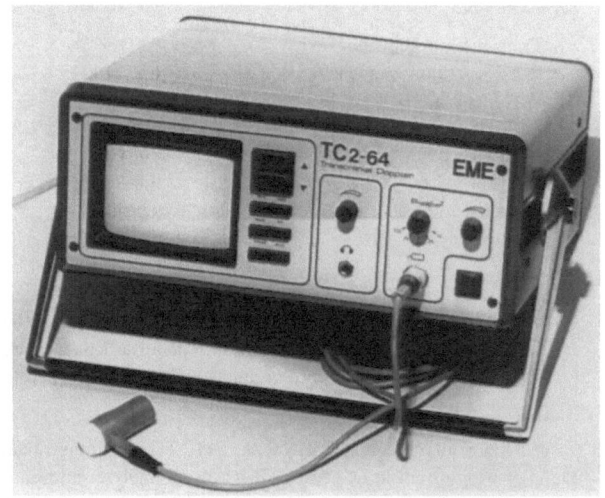

Fig. 48. The transcranial 2 MHz-Doppler with integrated frequency analyzer (EME, D-7770 Ueberlingen, Federal Republic of Germany)

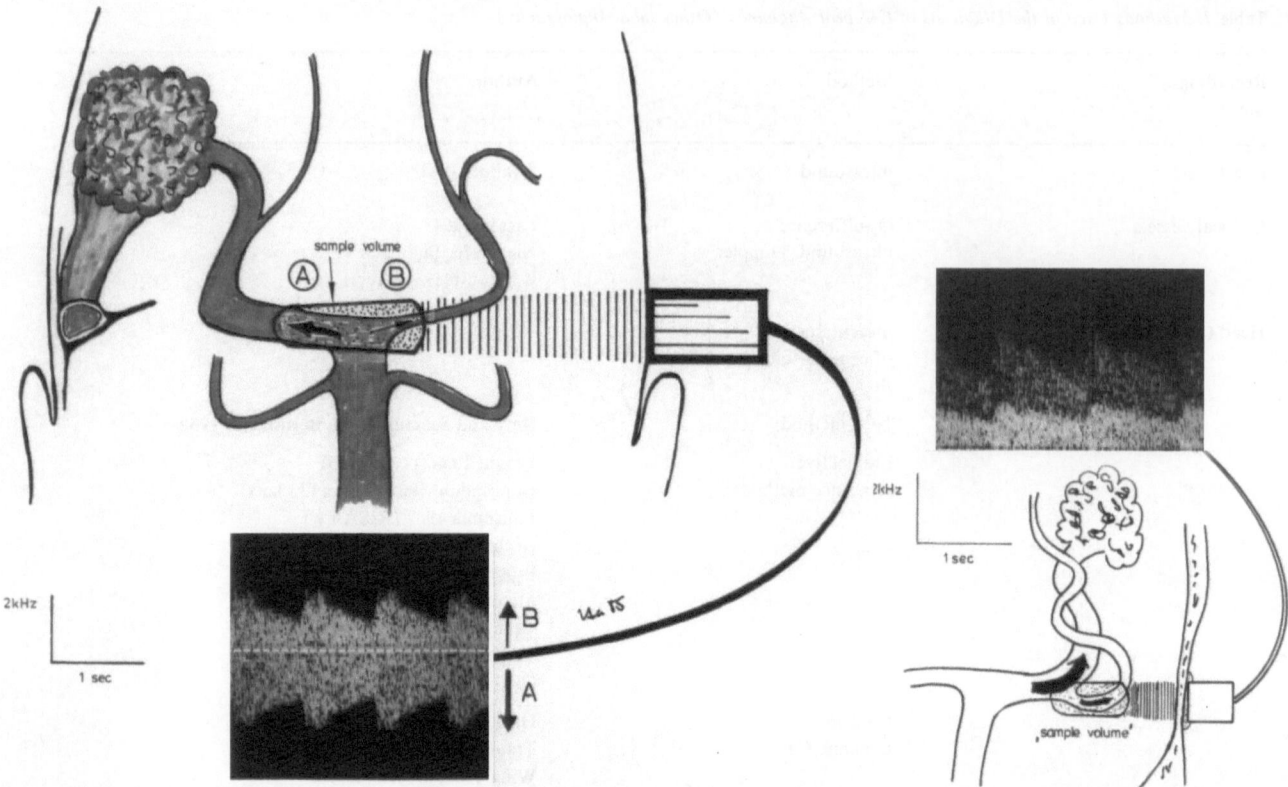

Fig. 49. Several vessels may lie within the sample volume. Left: Evaluation of the basilar bifurcation gives 2 signals of different flow direction. One belongs to the normal PCA (upward deflection), the other comes from the contralateral PCA (downward deflection). Right: Two MCA branches within the sample volume. The fast flow AVM feeder is evaluated at its marginal zone

Our studies were carried out with a prototype instrument constructed by Aaslid and Eden Medizinische Elektronik GmbH, D-7770 Ueberlingen, FRG, who kindly made the instrument available to us *(Fig. 48).

Technical Data

Transmission frequency	2 MHz
Pulse repetition frequency	6.8–18 kHz
Wave length	750 μm
Pulse duration	10 μsec
High pass filter	100 Hz
Low pass filter	3.4–9 kHz
Diameter of measuring probe	1.5 cm
Emitted energy	100 mW/cm

The sample volume of the emitted ultrasound beam is 1 × 0.5 × 0.5 cm, which is much larger than the vessel volume to be evaluated. Therefore, Aaslid developed an "acoustic lens" which focuses the beam in 5 cm depth to a sample volume of 0.9 × 0.4 cm

(Arnolds 1986). Nevertheless, this volume is greater than the basal arteries, so that small vessels in their proximity or tortuous parts of a vessel fall within the recording (Fig. 49).

With pulsed Doppler performance, recording depths can be shifted in steps of 0.5 cm from 3 to 8 cm, which is necessary for the evaluation of different basal vessels. The recorded frequency shift is displayed on a real-time frequency analyser (Angioscan); a flow adapter detects the direction of flow.

Aliasing

A problem occurring with pulsed Doppler techniques is the so-called aliasing. This means that flow patterns are divided when the frequency shift exceeds half the value of the pulse repetition frequency. This is why a 2 MHz pulsed Doppler device normally provides recordings only up to 6.25 kHz (250 cm/sec). With bidirectional spectral analysis, however, the Doppler frequencies exceeding this value are displayed below the baseline (Fig. 50). This is called aliasing (Phillips 1982). The divided patterns have to be added together; the upper recording limit thus reaches 12.5 kHz (500 cm/sec). Modern devices electronically correct the divided spectra (Fig. 50).

Relationship between Doppler shift, recording angle and flow velocity

The measured flow velocities depend on the angle of the insonation which is acute and lies between 0 and 30° for measurements from the MCA, ACA and PCA. The corresponding cosinus varies from 0 to 0.86; the maximal recording error is 14%.

---

* Now available commercially as the EME TC2-64 Transcranial Doppler with a built-in 64-point FFT spectrum analyzer and anti-aliasing device. Maximum measuring depth is now nominally 14.5 cm to permit examination of the contralateral vessels. Mean velocity in cm/sec is displayed on the monitor.

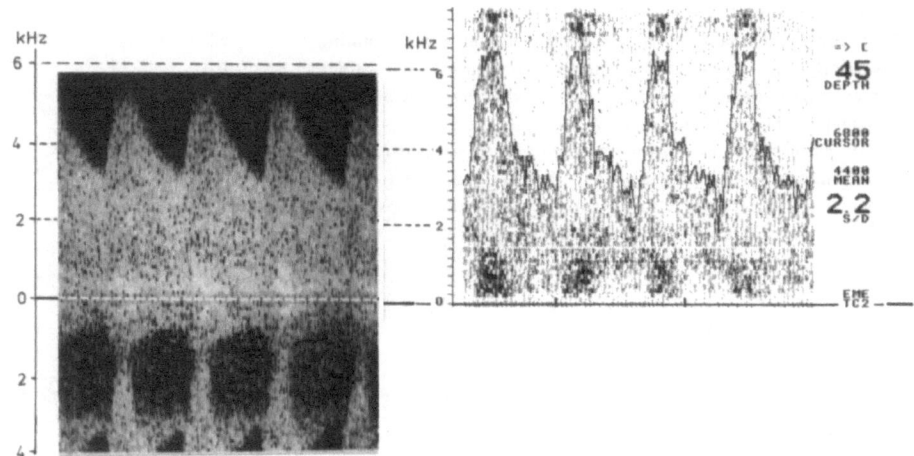

Fig. 50. Recordings in high flow velocities. Left: old device with divided display. Right: TC-2 device

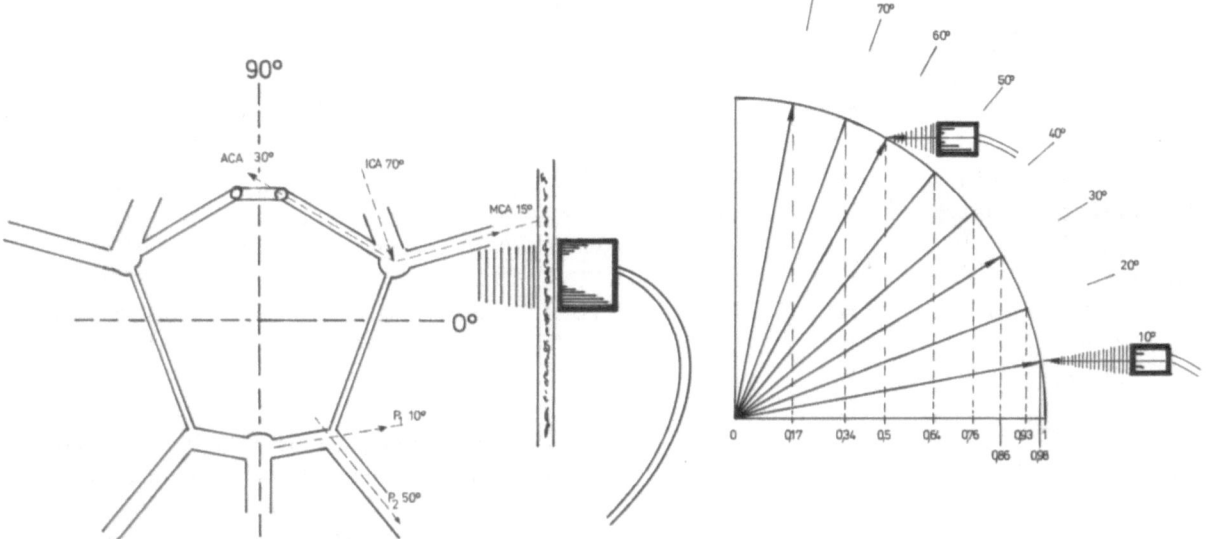

Fig. 51. Recording angles in transcranial Doppler sonography. The indicated degrees are average values obtained from 20 cranial CT evaluations. Doppler equation: see Chapter II.2.5.

$$f = v \times \frac{2\,fo}{c} \times \cos\alpha$$

f = Doppler shift in Hz, v = blood flow velocity, fo = mean values of emitted frequency, c = ultrasound velocity in tissue (cm/sec), α = angle between the direction of emitted sound beam and blood flow

Usually, flow velocities are indicated in kHz of Doppler shift. As these values depend on the transmission frequency of the device (2–20 MHz), the recorded frequencies are often not comparable. This can be overcome by expressing velocity in cm/sec (Aaslid). For the 2 MHz device, one kHz of Doppler shift corresponds to a flow velocity of 39 cm/sec.

Recording technique

The basal cerebral arteries can be identified according to depth and direction of the recording, the measured flow direction, the frequency pattern and to the results of compression tests of the CCA.
— Identification of MCA:
The recording depth is between 2.5 and 5 cm; flow direction is usually towards the probe. Compression of the ipsilateral CCA results in a flow reduction; "stuttering" compression leads to a wave-like recording pattern (Arnolds 1986).
— Identification of ICA:
The recording depth is 5–5.5 cm; the flow direction is towards the probe which points slightly basal and anteriorly. Compression of ipsilateral CCA will show a noticeable reduction, a standstill or reversed direction of blood flow in the ICA.
— Identification of A1:
The recording depth is 5–6 cm with the flow directed away from the probe. The ultrasound beam points frontobasally. Compression of the ipsilateral CCA reverses the flow direction in the case of a patent anterior communicating artery; compression of the contralateral CCA will usually result in a two to threefold increase of flow velocity.

— Identification of PCA:

P1 becomes recordable in 5–6 cm depth with the probe pointing posteriorly and basally. Flow direction is towards the probe; compression of the ipsilateral CCA leads to increasing velocities due to collateral flow via the posterior communicating artery. Often, venous patterns are recordable coming from Vena basalis Rosenthal. P2 is found in 5.5–6 cm depth with its flow directed away from the probe. The frequency spectra are similar to that of P1; ipsilateral CCA compression leaves this pattern unchanged.

1.3. Normal Values

Aaslid reported his transcranial Doppler recordings in 50 healthy subjects. He discovered no significant inter-hemispheric differences and found the following mean flow velocities in basal cerebral arteries:

MCA	62 ± 12.0 cm/sec
ACA	51 ± 12.0 cm/sec
PCA	44 ± 11.0 cm/sec
ICA	37 ± 6.5 cm/sec

Arnolds (1985) reported his findings in 51 subjects (kHz): (1 kHz $\doteq$ 39 cm/sec)

MCA	systole	2.30 ± 0.30 kHz
	diastole	1.15 ± 0.25 kHz
ACA	systole	1.80 ± 0.35 kHz
	diastole	0.85 ± 0.22 kHz
PCA	systole	1.50 ± 0.29 kHz
	diastole	0.74 ± 0.18 kHz
BA	systole	1.45 ± 0.31 kHz
	diastole	0.72 ± 0.22 kHz

Own findings

We recorded the blood flow velocity of the MCA in 70 patients of different age. Investigations were performed with the patient in a seated position; endexpiratory pCO_2 was measured using an infrared CO_2 analyser (Normocap-R). The measurements were started after a 5 minute pause with pCO_2 values of 40 ± 2 mmHg.

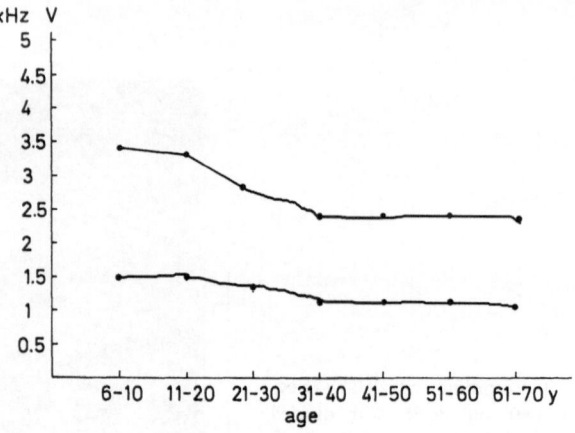

Age group	Number	Normal values (kHz)		pCO_2	Index of resistance R
		systole	diastole		
6–10	10	3.43 ± 0.2	1.5 ± 0.2	38.1	0.56
11–20	10	3.29 ± 0.5	1.49 ± 0.3	40.6	0.55
21–30	10	2.76 ± 0.3	1.3 ± 0.2	42.4	0.53
31–40	10	2.39 ± 0.2	1.11 ± 0.1	41.1	0.53
41–50	10	2.41 ± 0.4	1.13 ± 0.2	41.4	0.53
51–60	10	2.38 ± 0.3	1.1 ± 0.1	40.0	0.54
61–70	10	2.33 ± 0.1	1.0 ± 0.2	40.3	0.57

Fig. 52. Transcranially measured systolic and diastolic flow velocity in left MCA (recarding depth 4–4.5 cm) in different age groups

Transcranial recording depths in children were 4 cm, in adults 4.5 cm. Children below the age of 6 were not examined. Signals were obtained in all cases (Fig. 52).

The mean systolic MCA flow velocity of all subjects was 2.73 ± 0.34 kHz, the diastolic value was 1.24 ± 0.21 kHz (for more detailed data see Chapter IV.3).

Comparing the different age groups, the following observations were made (Fig. 52):

— systolic and diastolic MCA flow velocities decrease with age;

Table 8. *Percentage Change in Flow Velocity in MCA of Young (6–10 Years) and Old (61–70 Years) Age Groups*

	Age group 6–10 years n = 10		Age group 61–70 years n = 10		Percentage change	
	systole kHz	diastole kHz	systole kHz	diastole kHz	systole	diastole
MCA Normal Values	3.43	1.5	2.33	1.0	32%	33%

— the difference of systolic and diastolic amplitudes decrease with age;

— age-dependent decreases in MCA flow velocity are most pronounced up until the age of 40; almost no change occurs in later life (Table 8).

In all patients, the resistance index according to Pourcelot was calculated. No significant differences were found. The resistance index lies between 0.53 and 0.57 in all age groups.

Influence of blood pressure

In 10 healthy subjects between 20 and 30 years of age, MCA flow velocity was studied before and after application of a sympathomimetic drug. After 5 minutes of rest in the reclined position, the initial blood pressure was measured. Then, a 10 mg bolus of etilefrin (Effortil) was given intravenously. Blood pressure was measured automatically every 10 seconds; MCA velocities were recorded continuously. At the moment of highest blood pressure, flow velocities and resistance indices were documented and compared with the initial values. We found the following results (Table 9):

— all subjects showed changes in blood pressure, mean flow velocity and resistance index;

— the mean blood pressure increased by 25% and stayed within the limits of vascular autoregulation;

— MCA flow velocities increased by 14%;

— the resistance index increased by 21%.

Fig. 53 shows the continuous registration of an MCA pattern under rising blood pressure. Simultaneous changes of the resistance index and mean flow velocities are displayed on the left side. Initially, the peripheral vascular resistance increases more than the flow velocity. The first parts of the curves (1–4) show an opposite course of flow velocity and stream resistance, whereas their latter parts run in almost the same direction. In the end, the resistance index is lower than in the beginning, whereas flow velocity still remains elevated.

1.4. Discussion (Normal Values)

Transcranial Doppler sonography provides a useful method for the detection of blood flow velocities in basal cerebral arteries. The relation of maximal systolic and end-diastolic velocities reflects the peripheral stream resistance of brain arterioles. Thereby, valuable information on changes in intracranial hemodynamics can be obtained in a noninvasive fashion. Mainly depending on the unknown recording angle, the possible error rate is under 15% (Aaslid 1982).

We investigated 70 healthy subjects of all age groups. All measurements were performed in one artery (MCA) with the same recording depth. In each case, the strongest signal obtained was evaluated. In normal subjects, the maximal systolic flow velocity was 2.1 kHz

Table 9

Case	Age	Normal blood pressure				After 10 mg Effortil®			
		SAP mm Hg	SAP mean mm Hg	Time averaged peak velocity kHz	R	SAP mm Hg	SAP mean mm Hg	Time averaged peak velocity kHz	R
1	29 f	123/67	84.8	1.6	0.58	138/73	94.6	1.8	0.64
2	26 f	112/66	81.2	1.7	0.53	140/81	100.6	1.9	0.66
3	26 f	107/73	84.2	2.4	0.49	126/77	93.2	2.5	0.61
4	28 m	123/58	79.6	1.7	0.50	142/83	102.6	2.0	0.65
5	26 f	125/75	91.6	1.7	0.57	160/100	119.9	2.0	0.69
6	28 f	116/66	82.6	1.8	0.54	141/78	98.9	2.0	0.61
7	25 f	119/65	82.9	1.5	0.54	163/94	116.8	1.75	0.61
8	21 f	116/59	77.9	2.1	0.52	143/76	98.2	2.4	0.56
9	29 m	103/49	66.9	1.2	0.50	143/74	96.9	1.6	0.68
10	23 f	110/68	81.9	1.6	0.45	131/72	91.6	1.9	0.57
	26.1 ± 2.6		81.3 ± 6.2	1.71 ± 0.3	0.52 ± 0.03		101.3 ± 9.5	1.98 ± 0.2	0.63 ± 0.04

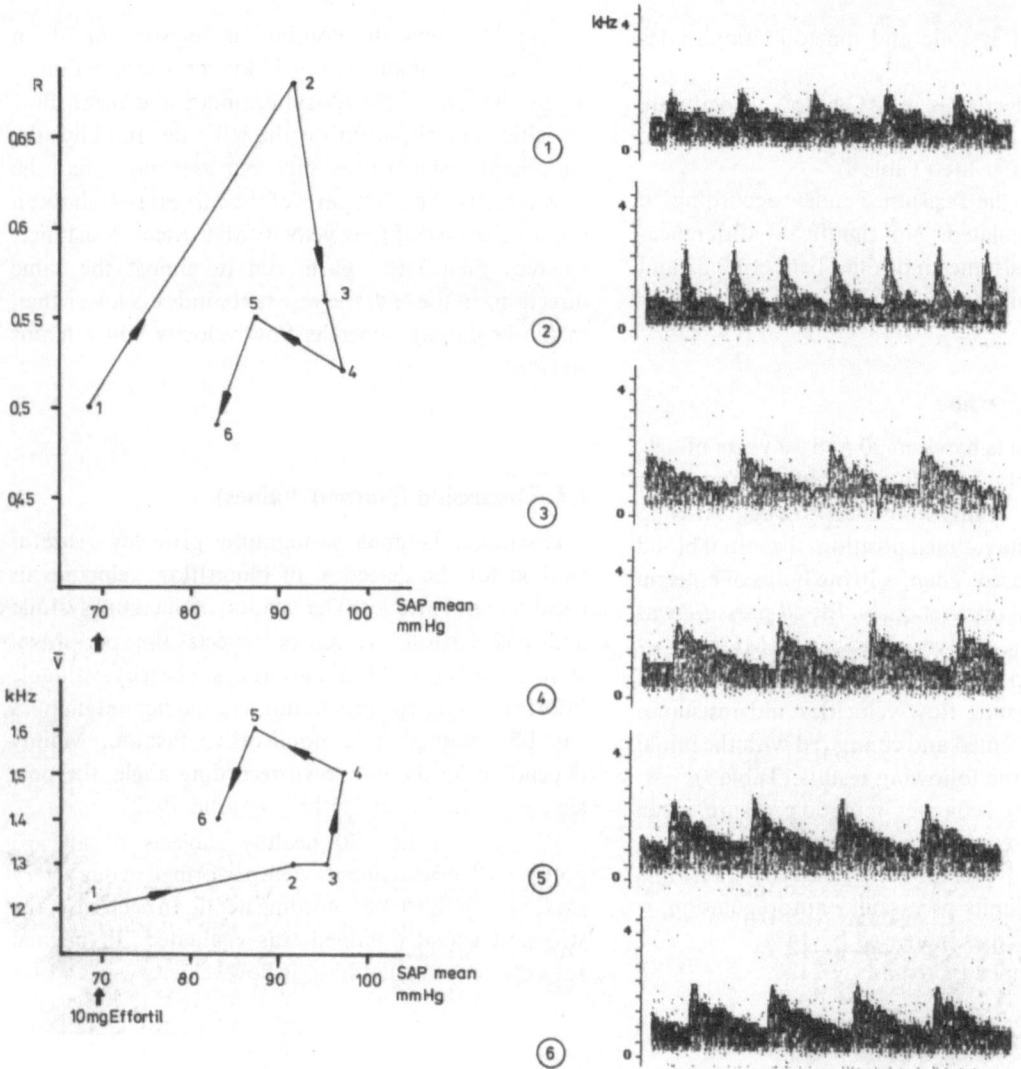

flow velocity, index of resistance in normal and high blood pressure

Fig. 53. Resistance index (R), mean flow velocity (v̄) and Doppler spectra after 10 mg of etilefrin

(82 cm/sec, 2.9 km/h). The highest velocities ever measured in humans were found in cerebral vasospasm and reached 420 cm/sec (15 km/h).

Systolic and diastolic velocities in the MCA decrease proportionally with age. Between the age groups of 6–10 years and 61–70 years, a flow deceleration of 32% (systolic) and 33% (diastolic) occurs. Similar results have been obtained by Arnold (1986) in 51 subjects. Former authors have reported on age-dependent velocity decreases in extracranial vessels (Mol 1974, Colon 1979). CBF measurements have shown decreasing brain perfusion in the elderly (Melamed 1980, Meyer 1978, Gündling 1985).

We found the resistance index of the MCA to remain stable throughout all age groups.

Sudden increases in systemic blood pressure within the limits of vascular autoregulation result in increasing MCA flow velocities and resistance indices. A 25% increase in mean blood pressure was followed by a 14% blood flow acceleration and a 21% rise in the resistance index. During these surprising changes the major rise in peripheral vascular resistance precedes that of flow velocity. Probably, the "delayed" flow acceleration reflects hyperemia following a short-term vasoconstriction.

2. Transcranial Doppler Sonography in Angioma Patients (Normocapnia)

2.1. Preoperative Recordings

2.1.1. Flow Characteristics of Angioma Supplying Arteries

Fifty patients with angiographically proven cerebral angiomas were studied. In each of them, all basal cerebral arteries were evaluated in varying recording depths from 3–6.5 cm. Systolic and diastolic flow velocities were measured and the resistance indices calculated. Recordings were performed after a 5 minute rest period. Systemic blood pressure and pCO_2 (infrared) were also documented.

Results

Pathological flow patterns were detectable in all large angiomas. Number and flow direction of the main feeding arteries (MCA, ACA, PCA) could be determined. Altered blood flow distribution and steal effects were discernable. Depending on the angioma's location, some of the smaller malformations (below 2 ccm) could not be unequivocally identified.

Classification of angioma supplying arteries according to flow velocity and resistance index

The flow velocity and resistance index of a vessel vary depending on how much it contributes to an AVM or supplies normal brain. We set up the following classification:

a) Exclusively angioma supplying arteries (Fig. 54).

— High systolic and diastolic flow velocities;
— Systolic velocity above 4.5 kHz (180 cm/sec), diastolic velocity over 3.5 kHz (140 cm/sec).
— Small differences between systolic and enddiastolic flow velocities with diastolic-systolic ratios being at least 74%.

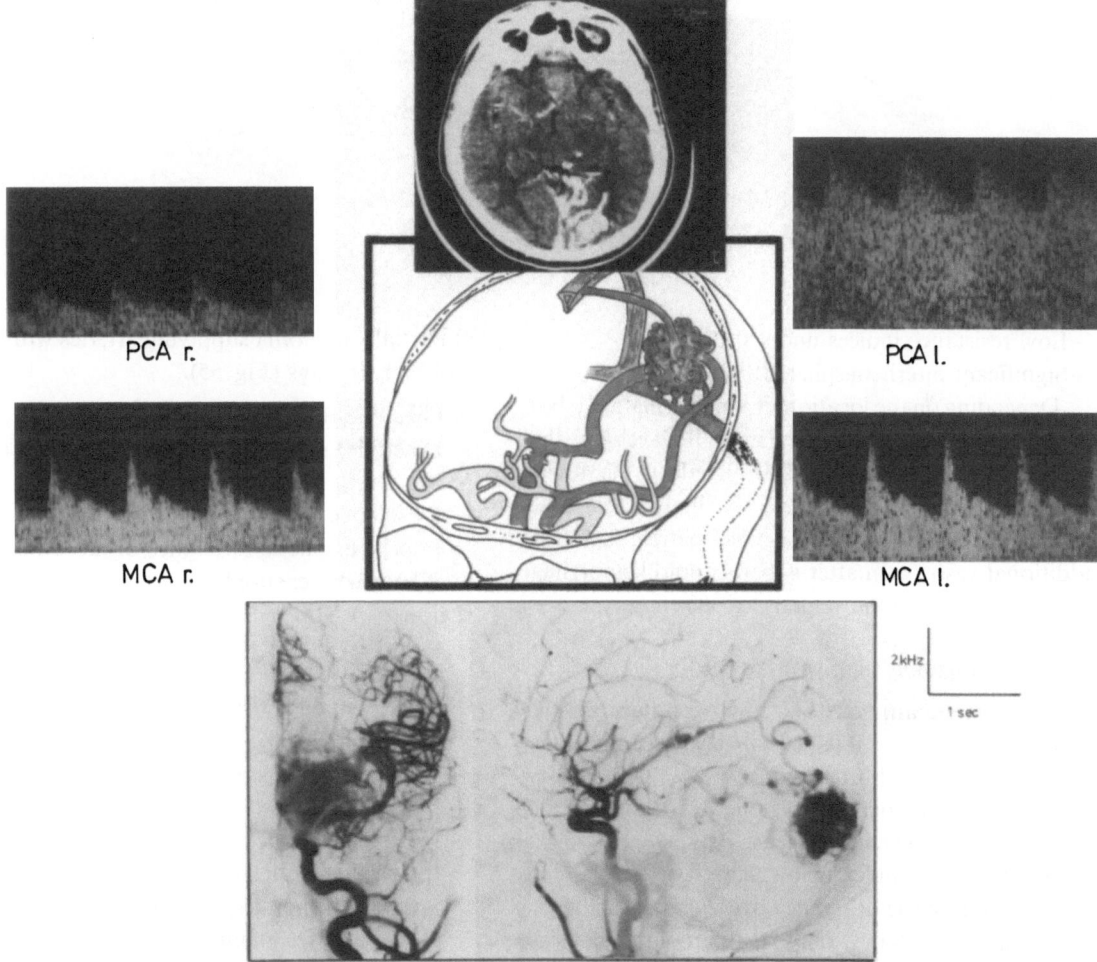

Fig. 54. Left occipital angioma: left PCA shows Doppler spectra typical for pure angioma feeders; left MCA is angiographically partially involved in the AVM supply and shows suspicious Doppler patterns

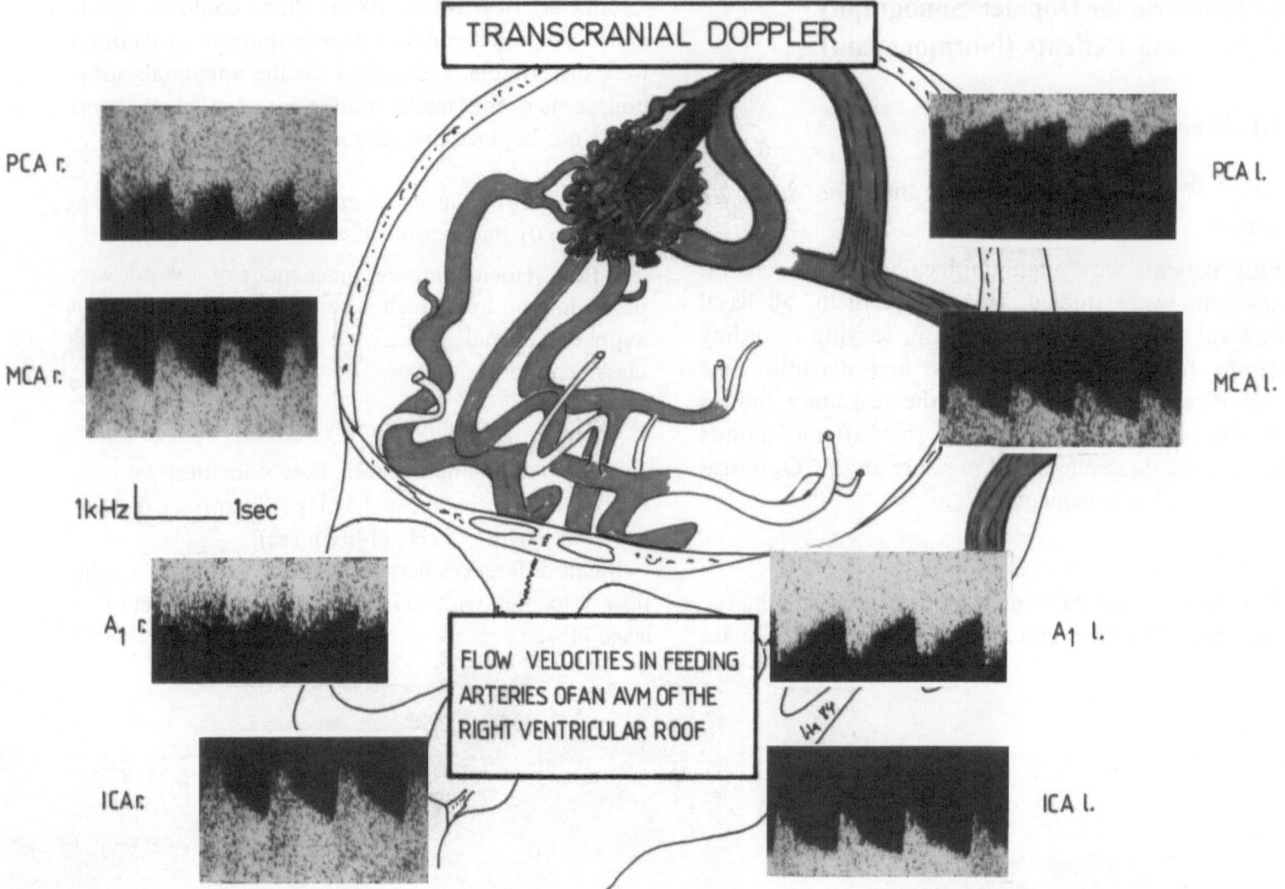

PCA as being mainly, MCA and ACA as being partially AVM supplying vessels

—Low resistance indices under 0.27.

—Significant interhemispheric difference.

Depending on the location of a malformation (short feeder, long feeder), the highest mean flow velocity that we found was 278 cm/sec (short feeder), the lowest resistance index was 0.24 and the highest diastolic-systolic velocity ratio 76%. One patient suffering additional vasospasm after subarachnoid hemorrhage even showed a systolic velocity of 300 cm/sec.

b) Mainly angioma supplying arteries.

—High systolic and diastolic flow velocities; systolic velocity 3.5–4.5 kHz (140–180 cm/sec), diastolic velocity 3–3.5 kHz (120–140 cm/sec).

—Small difference between systolic and diastolic velocities with ratios reaching at least 70%.

—Low resistance index below 0.30.

—Interhemispheric difference over 1 kHz.

From these arteries, only small amounts of blood supply normal brain, whereas their main supply runs to the angioma (Fig. 55).

c) Partially angioma supplying arteries with suspicious Doppler findings (Fig. 55).

—Diastolic flow velocities over 2 kHz (80 cm/sec).

—Diastolic—systolic velocity ratios of more than 55%.

—Resistance index below 0.5.

—Interhemispheric difference of more than 0.5 kHz.

These arteries supply angioma and also larger brain regions.

d) Angioma supplying arteries with normal Doppler findings (Fig. 56).

—Normal flow velocity.

—Normal systolic-diastolic velocity ratio below 50%.

—Normal resistance index.

—Interhemispheric difference of more than 0.5 kHz between corresponding vessels.

These arteries mainly supply brain. Feeders of very small or distally located AVM may show such characteristics.

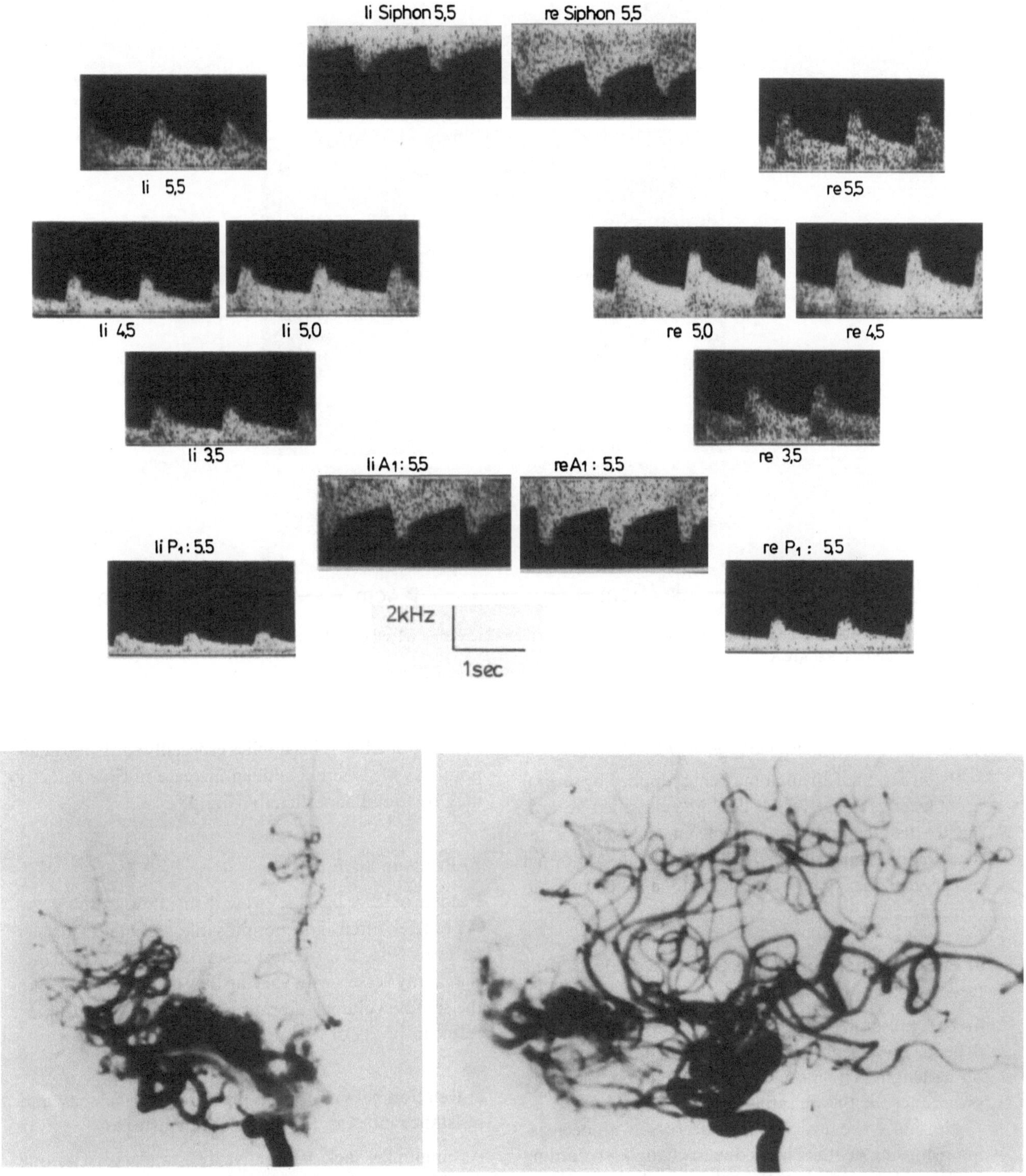

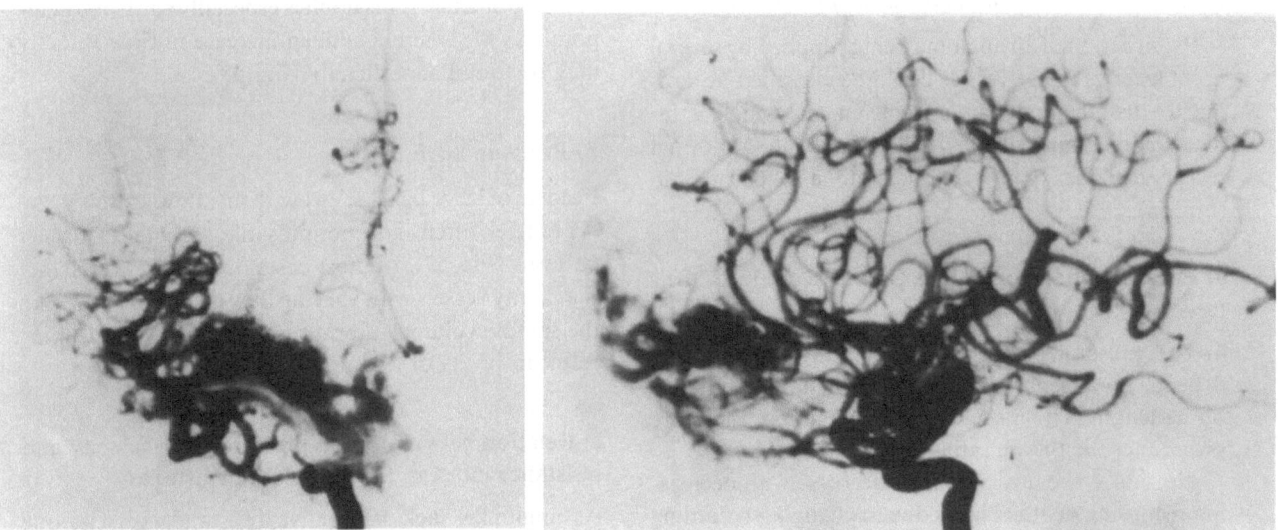

Fig. 56. Right frontobasal medial angioma: Right MCA and right ACA contribute to the angioma. Dopplersonographically, typical characteristics of AVM feeders are missing. The side difference exceeds 0.5 kHz in the MCA

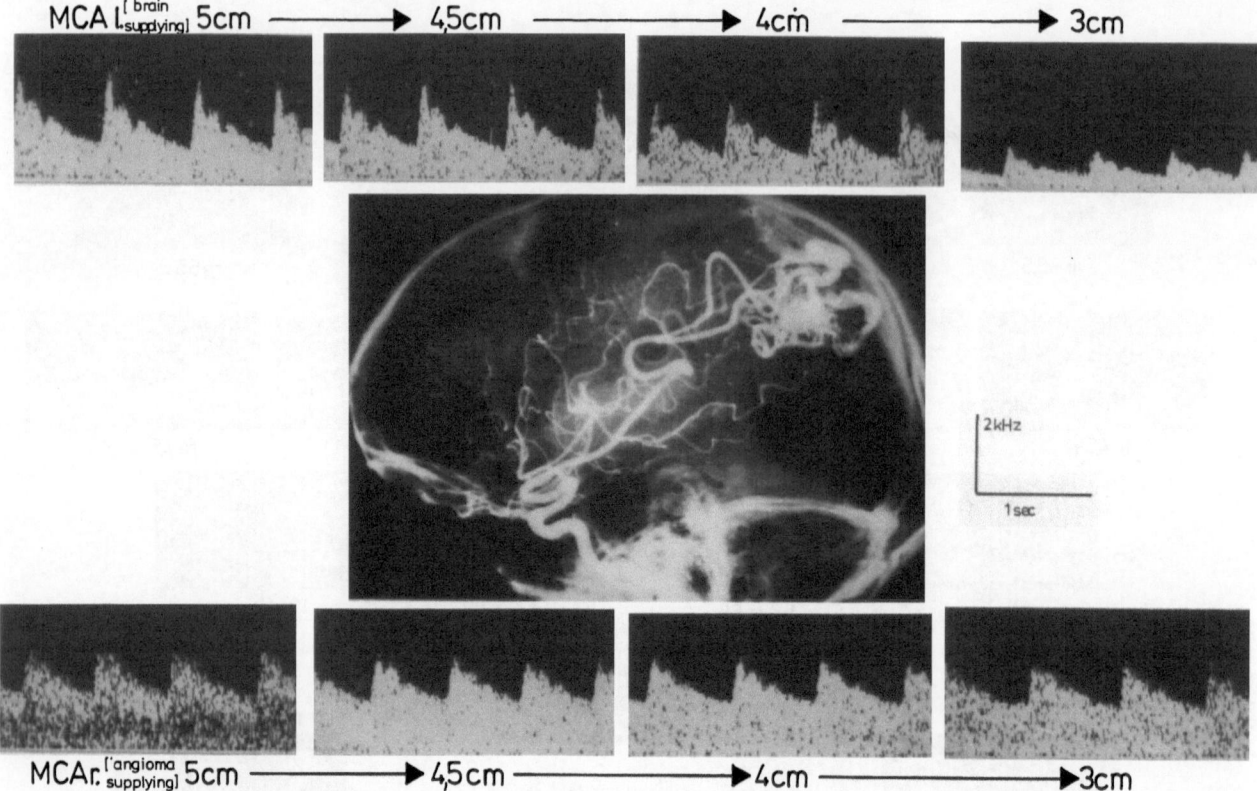

Fig. 57. Right parietal angioma: The AVM supplying right MCA shows increasing velocities with more superficial evaluation; normal velocity decrease in contralateral MCA

Other characteristics of angioma supplying arteries

In the situation of an angioma being supplied by several basal arteries, these vessels could be classified according to their hemodynamic involvement, which is indicated not as much by flow velocity as by the abnormal resistance index and diastolic-systolic velocity ratio (Fig. 55).

a) Location of angiomas and Doppler spectra of their feeders

— The pulsatility of the Doppler pattern not only depends on a feeder's degree of AVM supply, but also on its length. The more distal the angioma, the more pronounced is the pulsatility in its feeder.
— Normally, the flow velocities decrease with decreasing recording depth. This is due to changing recording angle and to the fact that the vessels branch out superficially. As a characteristic in angioma cases, flow velocities remain unchanged with decreasing recording depth (Fig. 57). In small angiomas with normal transcranial Doppler findings in the basal vessels, increased velocities can be detected in more superficial,

dilated branches. This applies especially for frontotemporal AVM, where a sudden increase in flow velocity may be found superficially (Fig. 58).

b) Bruits in angiomas

Feeders of large basal AVM with fast flow as well as the AVM itself often show nonpulsatile, turbulent Doppler patterns (Fig. 59). In most cases, this is due to the fact that many vessels with varying flow directions lie within the sample volume. Vibrations of the vessel walls may result in bruits of about 800 Hz.

c) Relation between blood pressure, flow velocity and resistance index in angioma feeding arteries

As angiomas lack vasoactive resistance vessels, one would assume that flow velocities change in proportion to changes of the systemic blood pressure (BP). With higher BP, velocities should increase, whereas lower BP should result in flow decelerations.

In three MCA supplied angiomas, BP changes were induced preoperatively with the patient already under

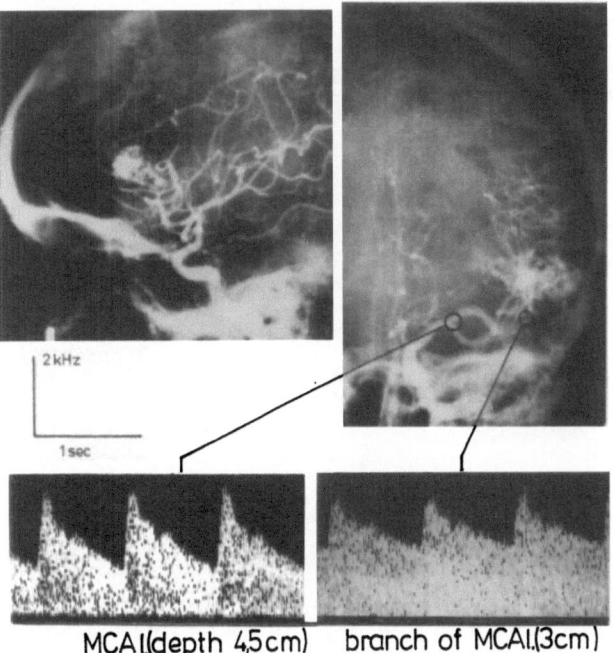

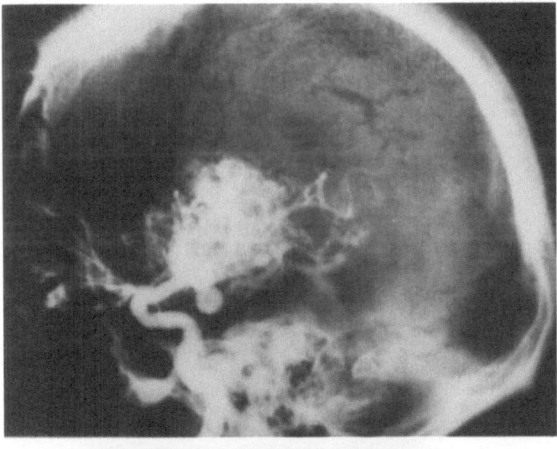

MCAl(depth 4,5cm)   branch of MCAl.(3cm)

Fig. 58. Small left frontolateral angioma: Normal spectra in the MCA's main trunk; typical AVM pattern in more superficial AVM-feeding branch

Fig. 59. Large left Sylvian angioma: Turbulent, nonpulsatile flow in the proximal MCA (depth 4.0 cm); bruits below 1 kHz

general anesthesia (Isoflurane⸌, causing no changes of autoregulation).

Results:
— Systolic flow velocities of MCA increase with rising BP and decrease with falling BP.
— Diastolic flow velocities do not increase with rising BP, but drop with falling BP.
— Mean flow velocities increase with rising BP and decrease with falling BP.
— The initially low resistance index shows a delayed increase under rising BP. It then drops only slightly with decreasing BP.

These investigations show that angiomas are indeed perfused in a pressure-dependent fashion. It remains unclear why the stream resistance becomes elevated in systemic hypotension compared to the initial normotensive situation. This may be due to dilatation of brain arterioles and consequently increased intracranial pressure.

2.1.2. Discrimination of Spastic and Angioma Feeding Arteries

Both angioma feeders and spastic arteries show high flow velocities. The highest value ever measured was 420 cm/sec (Aaslid 1985, personal communication). Differentiation between these conditions is impossible from recordings from one vessel alone. In general, the resistance index is normal, the systolic peaks are steeper and the heart rate is higher in vasospasm. The assumption of strictly unilateral flow acceleration occurring only in AVM is incorrect, as vasospasm may also appear in one vessel alone (i.e., sylvian hemorrhage). Angioma feeders may develop vasospasm following subarachnoid hemorrhage (Fig. 61).

2.1.3. Transcranial Doppler Sonographic Recordings and Angiographical Vessel Diameter

The flow velocity of a feeder is not the only criterion to assess the hemodynamic significance of an AVM. Flow velocities may be lower when several arteries contribute to the angioma. Theoretically, one could expect the velocity to also depend on the pressure gradient, the shunt volume and the size of the malformation.

The pressure gradient should be the same in small angiomas with large shunting vessels as in large malformations with small shunting vessels. Frictional forces also influence the blood flow velocity. They should be low in wide vessels and high in long ones. According to this, Nornes (1980) found the flow velocity to decrease

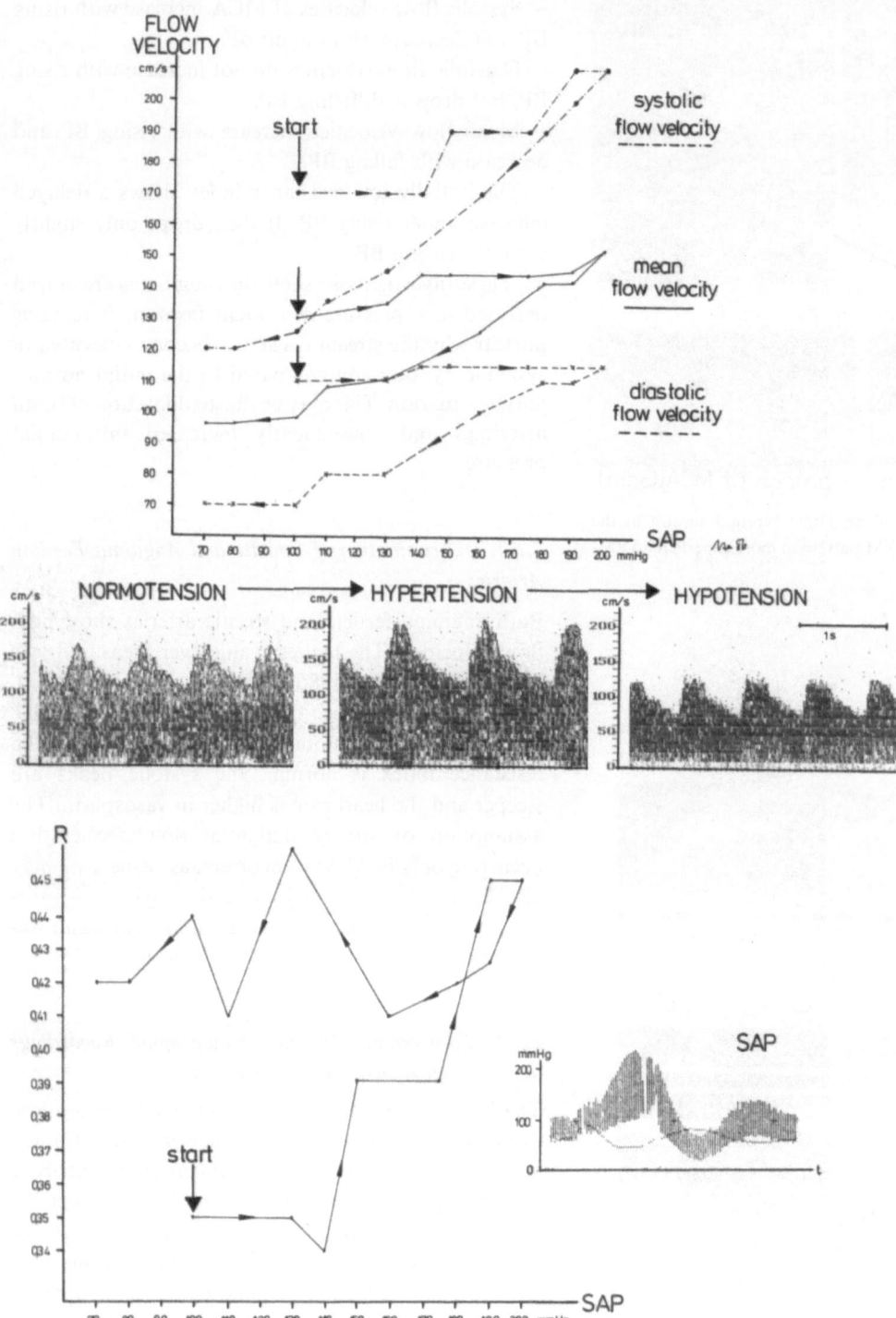

Fig. 60. Flow velocity (above) and resistance index (below) in angioma feeder (left MCA) under varying blood pressure

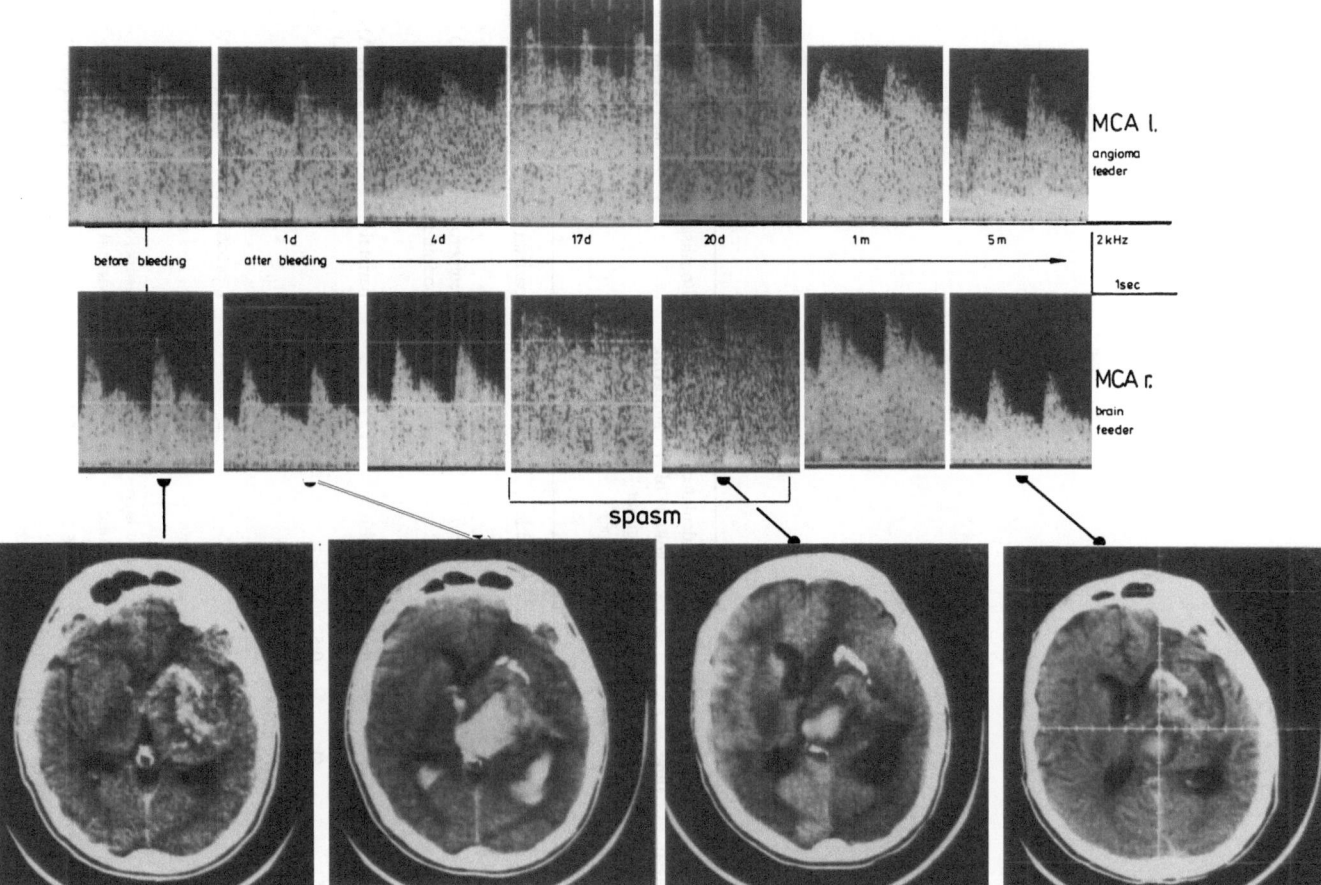

Fig. 61. Flow velocities in AVM-feeding left MCA and brain supplying right MCA before and after hemorrhage: vasospasm in both vessels after 17 days that subsides in the follow-up

with increasing length of feeders. Our findings suggest that besides length, the feeder's diameter strongly correlates with the flow velocity in angioma patients.

Parameters influencing blood flow velocity in angiomas (theoretical considerations)

— Systemic blood pressure
— Cross-sectional plane of feeders
— Shunt volume
— Angioma size
— Number of feeders
— Length of feeders

In order to verify some of these considerations, we compared our transcranial Doppler findings with angiographic results.

Method

Twelve patients with angiomas of varying size were studied by transcranial Doppler sonography prior to angiography. Mean flow velocities and resistance indices of all basal cerebral arteries were documented.

Angiographically, the real intravascular diameters (2 r minus röntgenographic magnification) and the vessel's cross-sectional planes (πr^2) were calculated. Number and length of feeding arteries as well as angioma volume were also measured. The flow rate of a given artery was then estimated using the following formula (V-mean in cm/sec, FV in ml/min).

$$FV = \pi r^2 \times V\text{-mean} \times 60.$$

The calculated flow rate FV differs from the real flow rate due to the unknown exact mean flow velocity in the vessel. Velocities measured by Doppler sonography mainly reflect the maximal values occurring in the center of the parabolic flow profile. The real flow velocity is lower. The error resulting from this can be disregarded when flow rates are compared interhemispherically in one patient and when the reference values from a healthy control group were obtained with exactly the same method.

The radiological intravascular mean diameters of ACA, MCA and PCA are known (Ring 1967, Wollschläger 1966, Gabrielson 1970)*. Together with the normal dopplersonographic mean values

---

* We used the values given by Ring 1967. They correlated best with our own measurements. The other authors report relatively large diameters and do not indicate any magnification factors.

Table 10

Case	f/m	Age	Vessels		Diameter 2 r cm	Length of angioma supplying vessel (cm)	Time averaged *peak* velocities (cm/sec)	Calculated flow volume (ml/min)	R	Calculated volume of angioma (cm³)	Number of supplying vessels	Localisation	Relation to the normal contralateral flow volume
			angioma supplying	brain supplying									
1 b+	m	46	MCA r.	+	0.324	16	117	579	0.28	3.5	1	parietal r.	102% (MCA)
			—	MCA l.	0.243		58.5	163	0.54				
2 b—	f	45	PCA r.	(+)	0.283	5	121	458	0.28	3.7	1	tempero-occipital r.	107% (PCA)
			—	PCA l.	0.226		58.5	142	0.55				
			—	MCA r.	0.251		78	232	0.52				
3 b+	m	36	MCA r.	+	0.251	16	85.8	255	0.44	4.1	3	parietal r.	57% (MCA)
			PCA r.	(+)	0.202	7	54.6	105	0.42				40% (PCA)
			ACA r.	(+)	0.121	15	58.5	41	0.5				
			—	MCA l.	0.251		54.6	162	0.65				
			—	PCA l.	0.162		39	48	0.67				
4 b—	f	26	MCA l.	(+)	0.34	14	105.3	574	0.5	5	2	l. precentral	23% (MCA)
			ACA l.	+	0.17	10	54.6	74	0.5				
			—	MCA r.	0.251		85.8	255	0.58				
5 b—	m	32	MCA l.	—	0.32	8	128.7	621	0.33	one large fistula	1	tempero-parietal l.	
6 b—	m	35	PCA l.	—	0.277	5	97.5	354	0.3	16.2	1	tempero-occipital l.	100% (PCA)
			—	PCA r.	0.231		48.7	123	0.33				

No.		sex	age												location	
7 b⌐	m	26	PCA l. MCA l. — —	— + MCA r. PCA r.	0.356 0.243 0.243 0.162	5 18	152,1 101.4 70.2 37.1	910 282 195 46	0.32 0.41 0.57 0.64	20.8	2	occipital l.	310% (PCA) 44% (MCA)
8 b⌐	f	32	PCA r. — MCA r. — ACA r.	+ PCA l. + MCA l. +	0.259 0.162 0.267 0.267 0.251	5 17 16	128.7 46.8 113.1 58.5 85.5	407 58 380 197 255	0.36 0.58 0.46 0.52 0.31	27	3	trigonum r.	175% (PCA) 93% (MCA)
9 b⌐	m	41	MCA r. ACA r. ACA l. —	+ — (+) MCA l.	0.275 0.405 0.356 0.234	7 1 1	79 136.5 113.1 54.6	279 1,168 677 142	0.4 0.22 0.2 0.4	86.5	3	fronto- basal medial	47% (MCA)
10 b⌐	m	44	ACA l. PCA l. MCA l. — steal	(+) — — MCA r. ← ACA r.	0.348 0.324 0.405 0.259 0.340	18 4 16	74.1 159.9 144.3 50.7 128.7	423 791 1,115 160 702	0.42 0.23 0.24 0.41 0.35	114	3	trigonum l.	185% (MCA)
11 b⌐	m	32	MCA l. PCA l.	—	0.421	2	156	1,303	0.31	154	2	sylvian l.	
12 b++	m	25	MCA l. PCA l. ACA r. — —	— — → steal + PCA r. MCA r.	0.299 0.17 0.210 0.121 0.218	2 3	144.3 132.6 93.6 46.8 54.6	610 181 195 32 123	0.43 0.41 0.51 0.63 0.62	155	2	sylvian l.	164% (MCA) 183% (PCA)

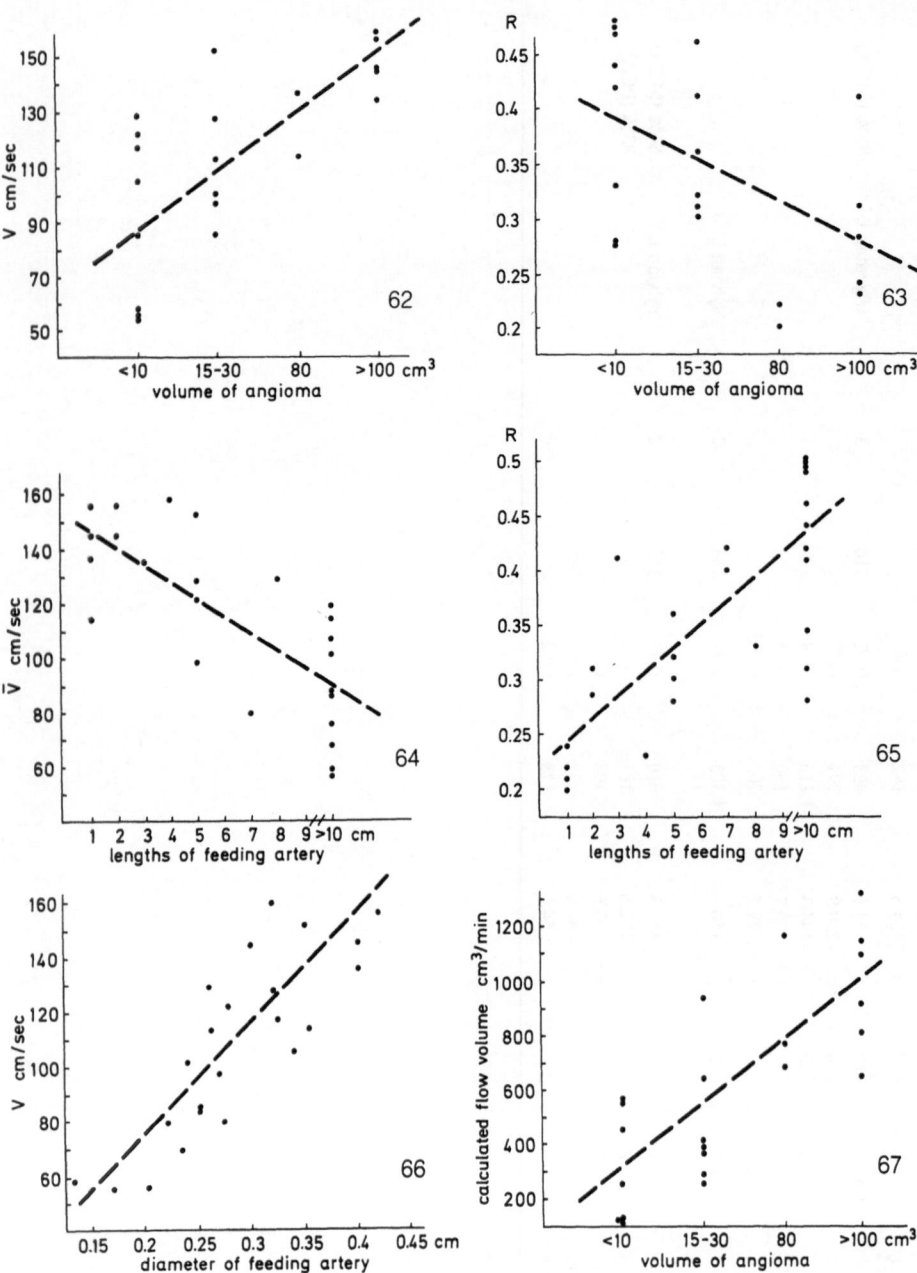

Fig. 62. Linear regression: Correlation between mean flow velocity of feeders and angioma volume (significant on 5% level in student's T-test, r = 0.508)

Fig. 63. Correlation between resistance index of AVM feeder and angioma volume (r = 0.508)

Fig. 64. Correlation between mean flow velocity and length of AVM-feeding vessels (r = 0.75)

Fig. 65. Correlation between resistance index and length of AVM feeders (r = 0.77)

Fig. 66. Correlation between calculated flow rate of AVM feeders and angioma volume (r = 0.79)

Fig. 67. Correlation between mean flow velocity and diameter of AVM feeders (r = 0.82)

of flow velocities in these vessels (Aaslid 1982), the mean flow rates can be calculated according to the formula cited above. They are

MCA	142.6 ml/min
ACA	91.1 ml/min
PCA	67.9 ml/min

These values can be compared to our pathological findings in angioma patients given in Table 10. Flow rates in angioma feeders may reach twenty-fold values compared to normal brain arteries and increase with the angioma volume.

Figs. 62–67 show the correlations we found between angiography and Doppler sonography:

— There is a correlation between blood flow velocity and angioma volume (Fig. 62). In all large angiomas, very high flow velocities are found, whereas most small angiomas show lower values. In the case of a small AVM having large shunting vessels, the flow velocity is high.
— There is a correlation between angioma size and peripheral vascular resistance of its feeder. Feeders of large angiomas show lower resistance indices (Fig. 63).
— There is a correlation between the length of a feeder and its blood flow velocity. Short feeders show fast flow, whereas longer feeders show decreasing values (Fig. 64).

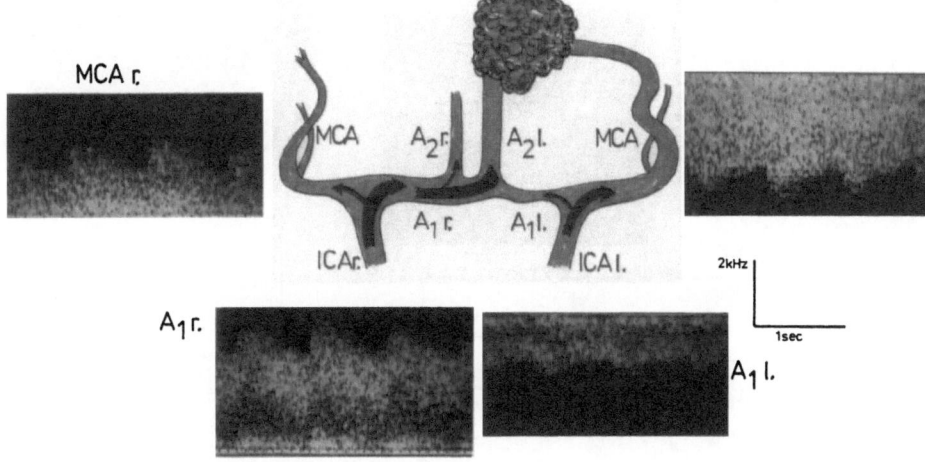

„STEAL" FROM A1 r. TO A2 l.
MCA r.
MCA
A2 r.
A2 l.
MCA
A1 r.
A1 l.
ICA r.
ICA l.
2kHz
1sec
A1 r.
A1 l.

Fig. 68. Altered blood distribution in large AVM of the left ventricular trigonum: The angioma-feeding left A2-segment is mainly supplied via right A1. Most of the left ICA flow is directed into the left MCA so that left A1 shows relatively low velocities

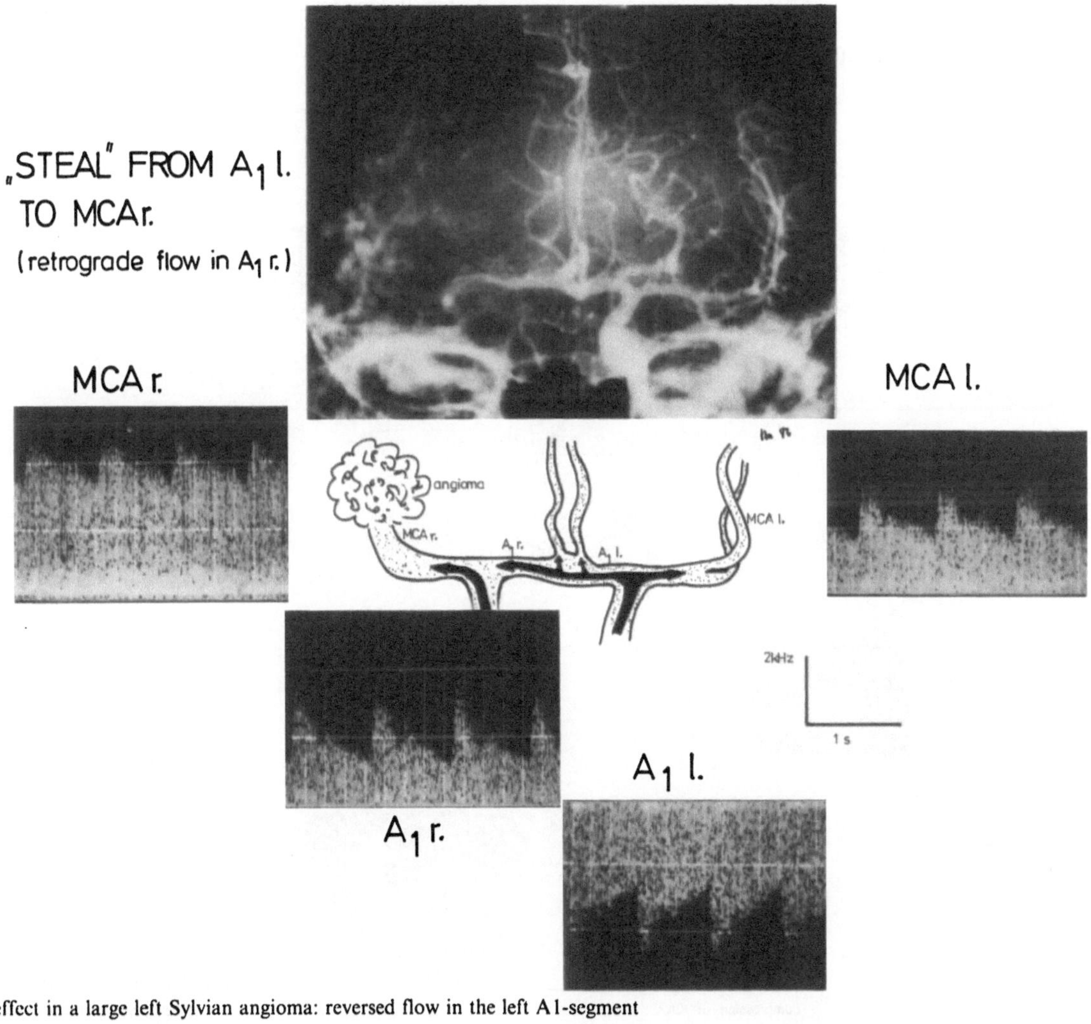

„STEAL" FROM A1 l.
TO MCA r.
(retrograde flow in A1 r.)
MCA r.
MCA l.
angioma
MCA r.
A1 r.
A1 l.
MCA l.
2kHz
1 s
A1 r.
A1 l.

Fig. 69. Steal effect in a large left Sylvian angioma: reversed flow in the left A1-segment

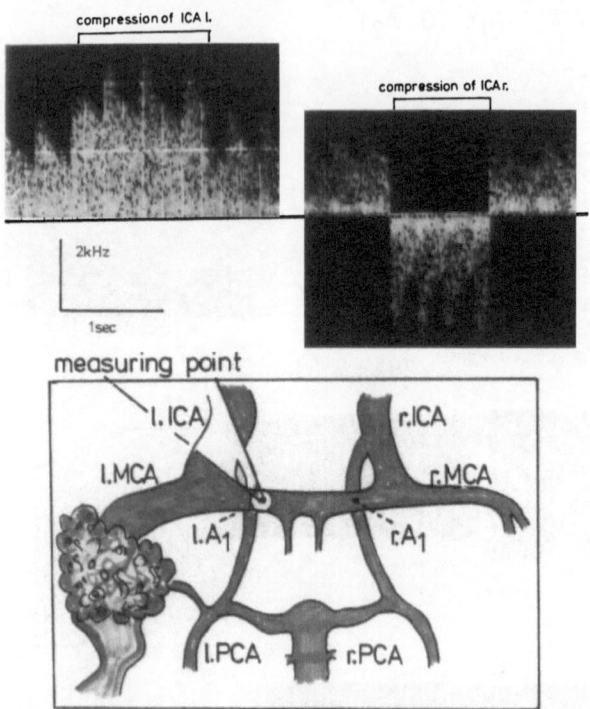

Fig. 70. Flow velocities in left A1 with compression of left and right CCA

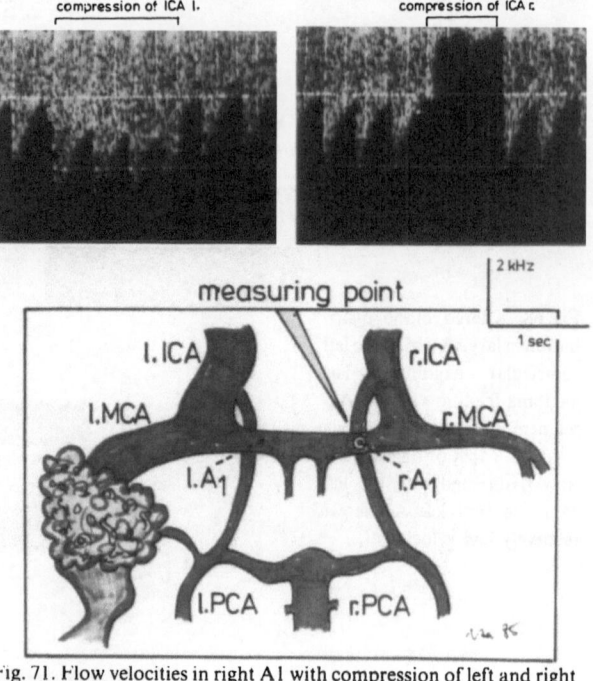

Fig. 71. Flow velocities in right A1 with compression of left and right CCA

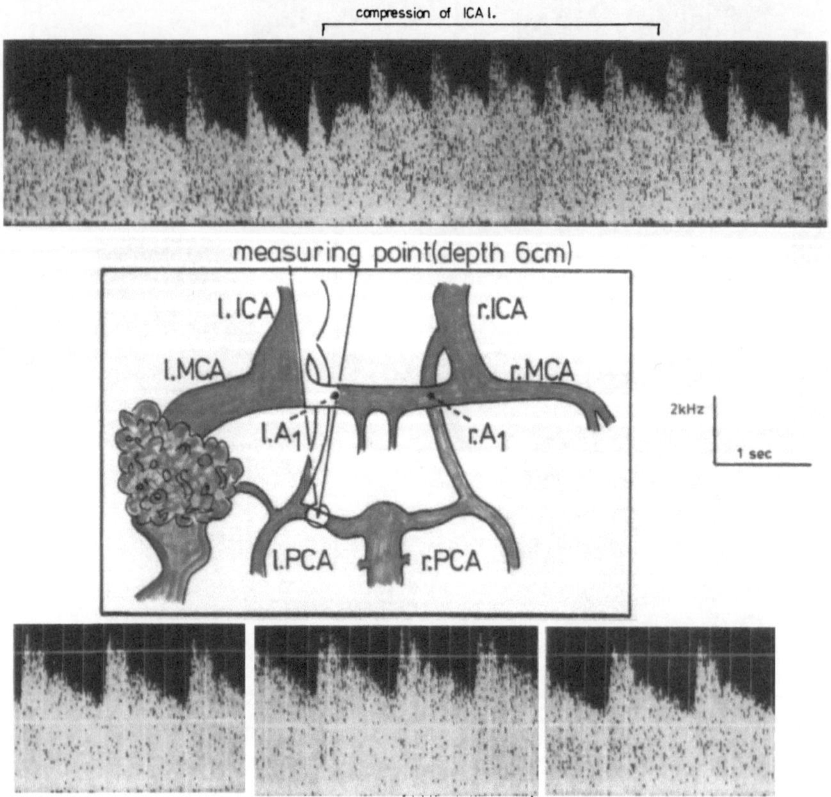

Fig. 72. Flow velocity in left PCA with compression of the left and right CCA

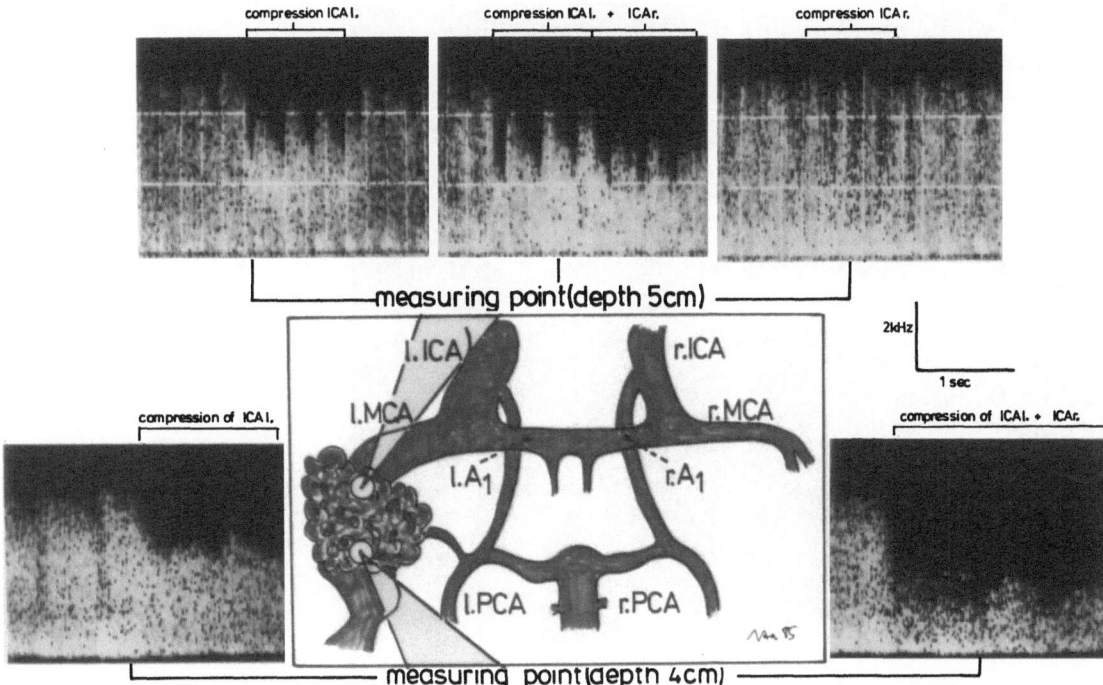

Fig. 73. Flow velocity of left MCA at its entrance into the AVM and of angioma outflow with compression of left, right and both CCA

— There is a correlation between length of a feeder and its resistance index. The longer the feeder, the higher its peripheral stream resistance (Fig. 65).
— There is a correlation between the diameter of a feeder and its flow velocity. The larger the diameter, the faster the blood flow (Fig. 66).
— There is a correlation between the calculated flow volume of feeders and the size of an angioma. The larger the angioma, the higher is the flow volume of the feeders (Fig. 67).

2.1.4. Altered Blood Distribution and Steal Effects

a) Blood distribution

Large AVMs angiographically show altered flow distributions. In large angiomas of the ventricular triganum for example, the angioma-feeding A 2-segment is mainly supplied via right A 1 (Fig. 68).

b) Steal phenomena

Steal effects occur when reversed blood flow in one or more cerebral arteries is induced by an angioma. Fig. 69 shows an example where reversed flow was found in the ipsilateral A1 in the presence of a large Sylvian AVM. Unequivocal identification of steal phenomena requires compression tests of the cervical common carotid artery (CCA).

2.1.5. Compression Tests in Angiomas

In angioma patients, information on the hemodynamic significance of the malformation ("sucking effect") can be obtained by compression tests of the cervical CCA. This is illustrated by the example of a large left Sylvian AVM (Fig. 70): Angiographically, the left carotid artery fills only the angioma feeding left MCA. The ACA is not visualized. The right carotid fills the right MCA, both A1s, left MCA and the angioma. With transcranial Doppler sonography, the right ACA shows anterograde flow. In the left ACA, reversed flow is indicated by the upward deflections of the frequency pattern (Fig. 70).

Transcranial Doppler evaluation of left A1 with CCA compression (Fig. 70):

With compression of the left CCA, flow velocities in the left A1 increase as a result of pronounced "steal" from the right carotid. If the right CCA is compressed, the flow in the left A1 turns anterograde.

Transcranial Doppler evaluation of right A1 with CCA compression (Fig. 71):

Compression of the left CCA causes accelerated flow in the right A1, which now mainly supplies the angioma. With compression of the right CCA, no signal is obtained from the right A 1, demonstrating insufficient crossflow from the left to the right anterior part of the Circle of Willis.

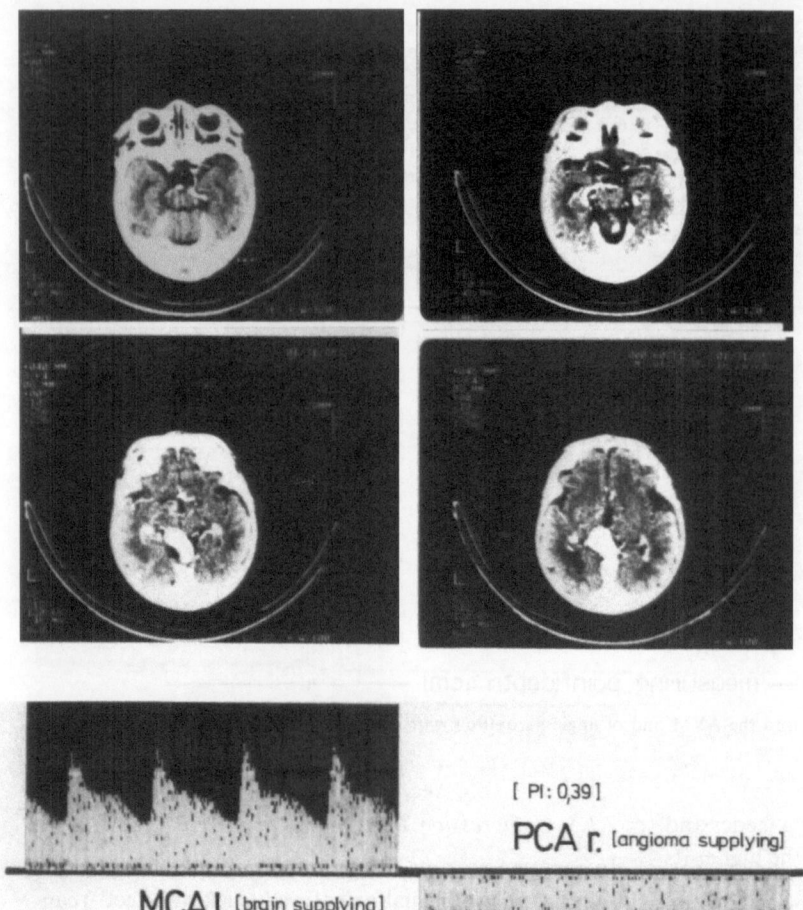

PCA r. [angioma supplying] [PI: 0,39]

MCA l. [brain supplying] [PI: 0,59]

2kHz

1sec

Fig. 74. Galenic angioma in a 10-day-old child suffering from cardial insufficiency with systemic hypotension: The brain-supplying left MCA shows elevated diastolic velocities indicating low stream resistance. The right PCA continues to show typical AVM patterns even under severe systemic hypotension (PI = resistance index)

Transcranial Doppler evaluation of left PCA with compression of CCA (Fig. 72):

With compression of the left CCA, the flow in the left PCA accelerates. Compression of the right CCA causes an only slight increase in flow velocity in the left PCA.

Transcranial Doppler evaluation of left MCA with CCA compression (Fig. 73):

With compression of the left CCA, the flow velocity in the main AVM feeding left MCA drops considerably. However, the Doppler pattern still shows the typical characteristics of angioma feeders with relatively high systolic and diastolic velocities. Even under bilateral CCA compression, diastolic flow velocities remain at 2 kHz. Compression of the right CCA does not influence the left MCA.

Transcranial Doppler evaluation of angioma outflow with CCA compression (Fig. 73):

The angioma's outflow shows high velocities with only little pulsatility. Compression of the left CCA reduces the velocity by 1/4. Compression of both CCA results in a further decrease to a now narrow, band-like frequency spectrum.

2.1.6. Special Cases

a) Angioma and hypotension

A ten-day old baby with a Galenic angioma also suffered from a patent Ductus Botalli leading to progressive heart failure with systemic hypotension. One day before death, the systemic blood pressure had fallen to 30/10 mmHg (Fig. 74). Transcranial Doppler evaluation of the feeding PCAs at this time still showed characteristic high diastolic flow velocities. Brain supplying MCA also had a relatively fast diastolic flow due to low peripheral stream resistance. Systolic velocities were the same in both vessels; the resistance index was 0.39 in the PCA and 0.59 in the MCA.

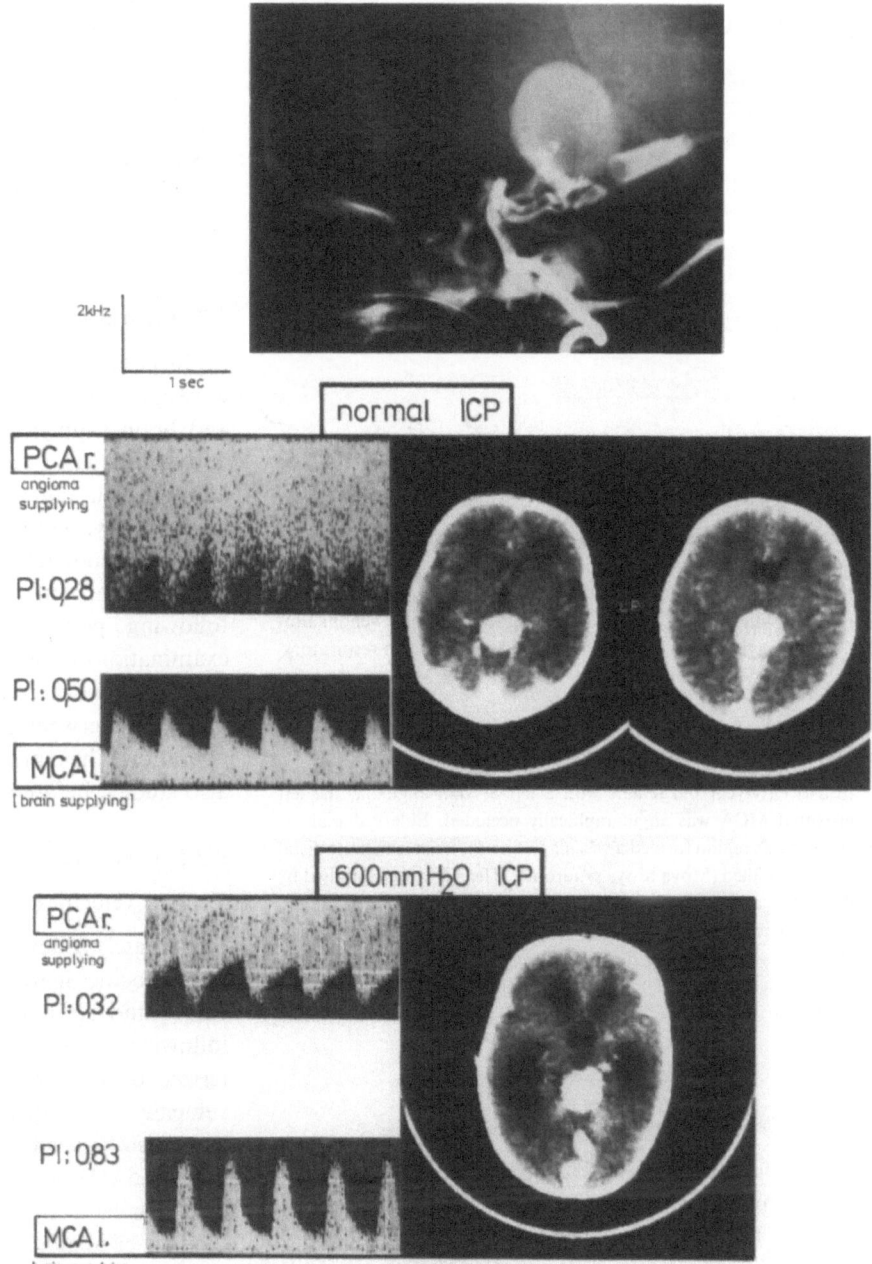

Fig. 75. Large galenic angioma: Flow velocities in the right PCA (AVM feeder) and left MCA (brain artery) under normal and raised intracranial pressure (*PI* resistance index)

b) Angioma and increased intracranial pressure

An eight-month old child had developed hydrocephalus after several bleedings from a large Galenic angioma. A few months later, it was readmitted because of shunt failure and presented with an intraventricular pressure of 60 cm H_2O. Transcranial Doppler recordings at this time suggested increased peripheral vascular resistance. Brain supplying MCA had decreased diastolic flow velocities and high, rounded-off systolic peaks. AVM supplying PCA still showed the typical features of an angioma feeder, though with decreased velocities and no more turbulencies. The calculated resistance indices demonstrate that even under increased intracranial pressure the peripheral vascular resistance of angioma feeders stays low (0.32),

whereas normal brain arteries show considerably increased peripheral resistance (0.83) (Fig. 75).

c) Angioma and vasospasm

Cerebral vasospasm can occur in angioma feeders. An angioma patient presented with intracerebral and subarachnoid hemorrhage. At the seventeenth posthemorrhagic day, flow velocities in the right MCA, which did not contribute to the AVM, were much higher than on admission. Angioma supplying right MCA also showed further flow acceleration to 7.5 kHz (see Fig. 61).

flow velocities of veins

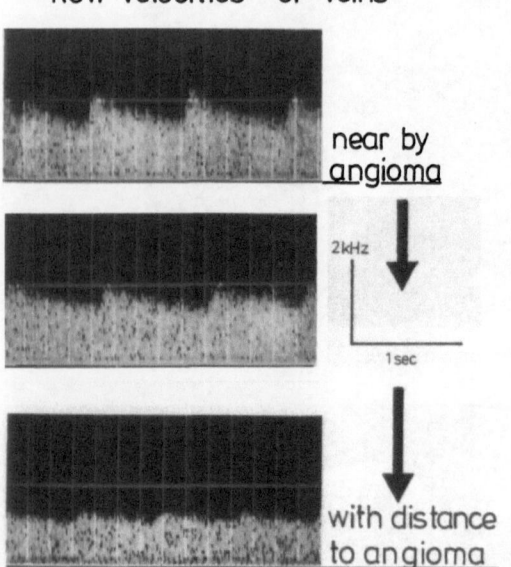

near by
angioma

2kHz

1sec

with distance
to angioma

Fig. 76. Flow patterns of AVM-draining vein: Pulsatile venous flow near the angioma levels off with increasing distance; no respiratory-dependent deflections occur

d) Angioma and Moya Moya syndrome

In a twenty-year old female with a left Sylvian angioma, the left proximal MCA was angiographically occluded. Enlarged pial arteries in this region formed a vascular network, from which the distal MCA was filled (Moya Moya syndrome). The AVM was supplied by this MCA and the ipsilateral PCA. Flow velocities of both internal carotid arteries were the same; left MCA also showed high values, although with good pulsatility. After surgical exclusion of the AVM, diastolic left MCA velocity remained unchanged, indicating low peripheral stream resistance in the presence of proximal MCA occlusion.

2.1.7. Flow Characteristics of Angioma Veins

Flow velocities of AVM draining veins can be detected transcranially. The patterns are characterized by flat systolic humps and a comparatively high diastolic amplitude. The closer to the angioma that a vein is evaluated, the more pronounced is its pulsatility. The normal, respiratory-dependent deflections due to rhythmic changes of intrathoracic pressure were never encountered in angioma veins. Venous velocities never exceeded 1 kHz.

2.1.8. Flow Characteristics of Cerebral Arteries

In general, cerebral arteries supplying the parenchymatous surroundings of angiomas show normal flow patterns. Only in cases of very large malformations may the diastolic amplitudes be relatively high compared to the systolic value (ratio above 50%). The resistance

indices are below 0.5. Such findings suggest slightly lowered peripheral vascular resistance in the surroundings of large AVMs. The best examples for this were occipital angiomas supplied by one PCA. In these cases, the basilar artery branches out in one AVM feeding and one brain supplying PCA. The latter usually show subnormal resistance indices.

2.2. Postoperative Recordings in Angioma Patients

In 30 patients, postoperative Doppler measurements were performed. We evaluated the former AVM feeders and brain supplying cerebral vessels; blood pressure and blood gases were measured simultaneously. In addition, all patients had immediate postoperative angiography. Transcranial Doppler sonography was performed immediately following surgery, then every 12 hours for three days and every 24 hours in the following period until discharge. The follow-up examinations were carried out after 3 and 6 months. In 2 patients, two-step surgery had been planned. Another patient underwent partial embolization prior to the operation. The remaining parts of these angiomas were also studied before complete removal.

2.2.1. Flow Characteristics in Former Angioma Feeders

a) Flow velocities after surgical AVM removal

Immediately following the removal of an angioma, the flow velocities in former feeders drop far below normal values and start to develop normal patterns in the following days (Fig. 77). The initial changes are characterized by very low, or even unmeasurable diastolic velocities. Early diastolic values are approximately 0, late diastolic velocities are slightly higher but remain below 0.5 kHz (20 cm/sec). These findings suggest very high peripheral stream resistance in the distribution of the former angioma feeders.

The time needed for subsequent normalization of the spectra depends on the size of the angioma and on the former degree of AVM supply by the vessel.

Stagnating arteries

For up to 9 days, former pure angioma feeders show high, steep systolic peaks on postoperative Doppler examinations. Diastolic velocities are very low or 0 (Fig. 78). These arteries become visible in postoperative angiography where they are characterized by their very large diameters and slow perfusion. Angiographic filling usually persists throughout the venous phase ("stagnating arteries", Hassler 1983). Intraoperative

Table 11

Case	Age	Vessels angioma	Vessels brain	Diameter 2 r (cm)	Time averaged peak velocities (cm/sec)	Calculated flow volume (ml/min)	R	Calculated volume of angioma (cm³)	Number of supplying vessels	Localisation	Course of treatment	%-change of flow volume	Total change of flow volume (%)
1a	40	MCA r.	(+)	0.275	79	279	0.4	86.5	3	fronto-basal medial	preoperative		
		ACA r.	—	0.40	136	1.168	0.22						
		ACA r.	(+)	0.356	113	677	0.2						
			MCA l.	0.234	55	142	0.4						
1b	40	MCA r.	(+)	0.283	85.5	325	0.33	17.9	3	removal of right part of angioma	after 1. operation	MCA r. + 16%	
		ACA r.	—	0.34	58.5	319	0.30					ACA r. −72%	
		ACA l.	(+)	0.356	117	700	0.32					ACA l. 3%	
			MCA l.	0.243	54.6	152	0.36					MCA l. + 7%	
1c	41	MCA r.	+	0.315	101.4	477	0.33	14.1	3	removal of right part of angioma	4 min later; before 2. Operation	MCA r. + 47%	
		ACA r.	—	0.40	58.5	452	0.38					ACA r. + 42%	
		ACA l.	—	0.461	85.8	862	0.31					ACA l. 23%	
			MCA l.	0.226	62.4	151	0.4					MCA l. −0.7%	
1d	41	—	MCA r.	0.243	39	108	0.85	0	0	total removal, large vessels	1 hour after angioma removal	MCA r. −77%	MCA r. −61%
		—	ACA r.	0.40	19.5	151	0.7					ACA r. −67%	ACA r. −61%
		—	ACA l.	0.445	19.5	182	0.83					ACA l. −79%	ACA l. −73%
		—	MCA l.	0.234	54.6	142	0.64					MCA l. −86%	MCA l. 0%
2a	26	PCA l.	—	0.356	152.1	910	0.32	20.8	2	l. occipital	preembolisation	PCA l. −34%	
		MCA l.	+	0.243	101.4	282	0.41					MCA l. + 15%	
		—	MCA r.	0.243	70.2	195	0.57					MCA r. 39%	
		—	PCA r.	0.162	37.0	45	0.64					PCA r. 80%	
2b	26	PCA l.	(+)	0.283	164	599	0.27	6.7	2	elimination of a large feeder of PCA	2 min after embolisation	PCA l. −84%	PCA l. −89%
		MCA l.	+	0.283	86	324	0.52					PCA r. −7%	PCA r. + 93%
		—	MCA r.	0.283	72	271	0.56						
		—	PCA r.	0.234	31	81.1	0.52						
2c	26	—	PCA l.	0.299	23.4	96.3	0.77	0	0	total removal of angioma	1 hour angioma removal		
		—	MCA l.	—	54.6	—	0.67						
		—	MCA r.	—	48.7	—	0.62						
		—	PCA r.	0.243	31.2	87	0.6						

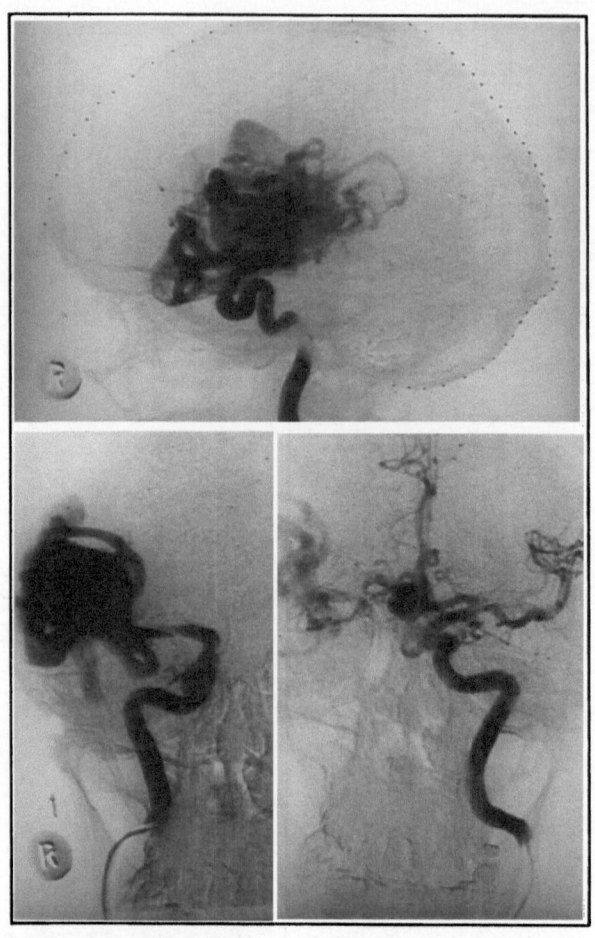

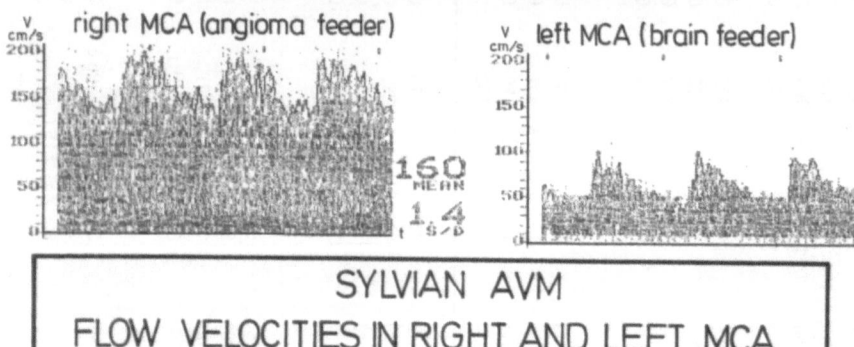

3. 77. Angiography and corre-
ınding TCJDoppler recordings
fore (a) and seven days after (b)
.noval of a right Sylvian AVM. I
am grateful to Professor Dr. Voigt,
Neuroradiology of the University
of Tübingen for the pre- and postoperative angiograms.

Fig. 77 a

measurements have demonstrated high intravascular pressure in such stagnating arteries when compared to the preoperative values in the same vessel. Intraoperative Doppler sonography has confirmed the transcranial findings showing high systolic peaks and low or no diastolic velocities.

Stagnating arteries have been demonstrated angiographically for up to the thirteenth postoperative day,

whereas the corresponding Doppler findings persist for up to 9 days (Fig. 87).

b) Flow velocities after partial exclusion of angiomas

In 2 patients, two-step surgery had been planned. Another patient underwent partial embolization of his AVM prior to surgery. Transcranial Doppler identifi-

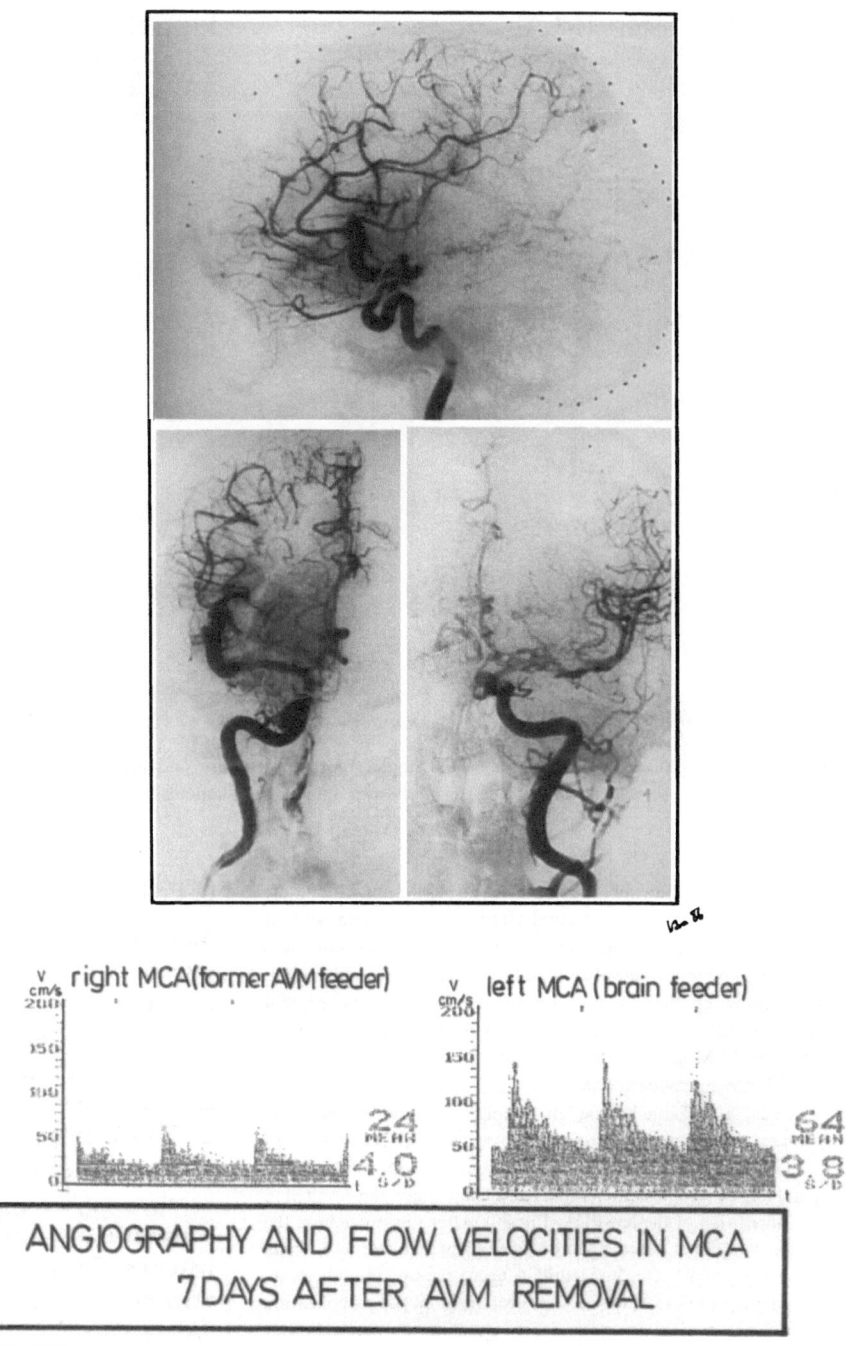

Fig. 77 b

cation of residual angioma parts was unequivocal in all 3 cases. The feeders' flow velocities had decreased but still showed abnormally high diastolic amplitudes and systolic-diastolic velocity ratios above 60% (Fig. 79, Table 11).

In case 1 of Table 11 the calculated volume of the large frontobasal medial angioma had decreased from 86 to 18 ccm after the first operation. Flow velocities in the main feeders had also dropped, leading to a 72% decrease in the calculated flow rate.

Vessels with partial contribution to the malformation showed slight increases of their calculated flow rates (16%, 3%), as did the exclusively brain supplying left MCA (7%). The feeders' resistance indices increased from 0.20 and 0.22 to 0.30 and 0.32 respectively. During the next four months, the angioma volume decreased further to 14 ccm, which must be attributed to thromboses in its marginal zone. The feeders' flow rates increased by 47%, 42% and 32%, whereas the brain-feeding left MCA remained unchanged. The left A1-segment became the main AVM feeder. The angioma's calculated flow rate increased from 677 to 860 ml/min, whereas the rate in its former main feeder (right ACA) dropped from 1,160 to 750 ml/min.

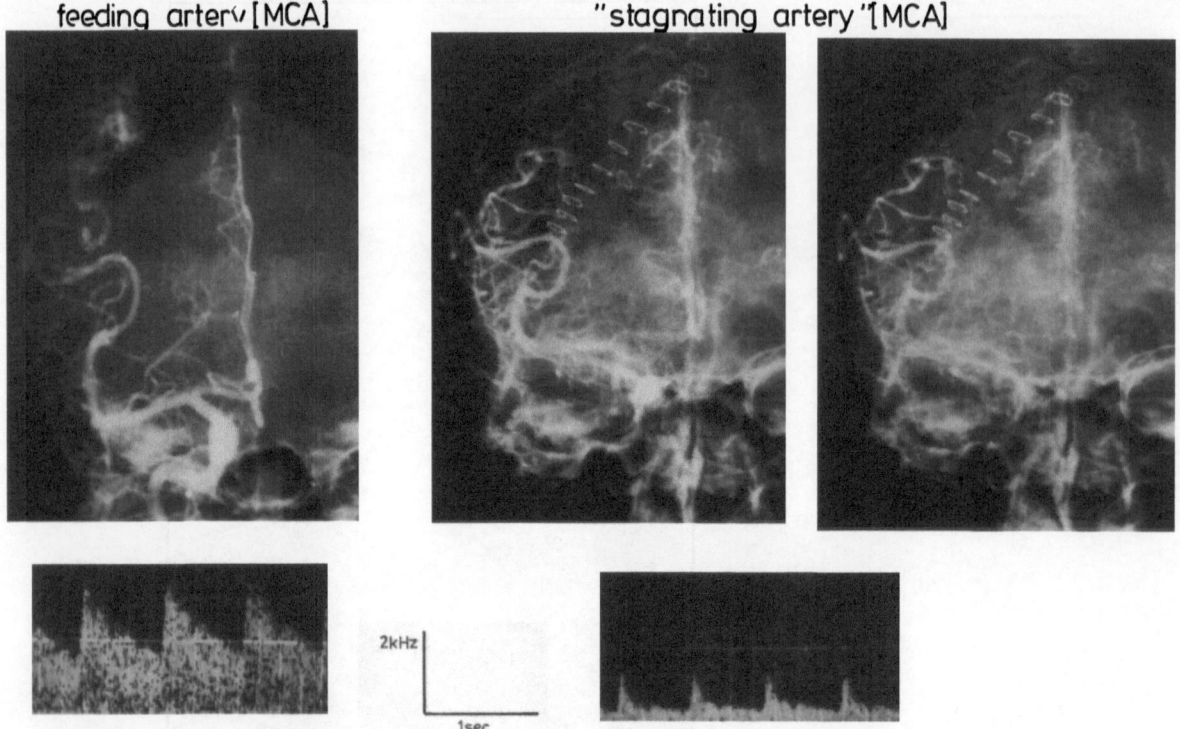

feeding artery [MCA]   "stagnating artery"[MCA]

2kHz

1sec

Fig. 78. Angiography and corresponding TC-Doppler recordings before and after removal of a right parietal AVM. The right MCA postoperatively appears as a "stagnating artery" showing unchanged diameter but very slow flow

Immediately following the complete AVM exclusion, blood flow rates dropped noticeably in cerebral arteries both with and without former contribution to the angioma. Peripheral stream resistance was elevated far above normal values for 5 to 6 days following surgery. On further follow-up, the flow rates were reduced to 39% of the initial value in partially AVM feeding right MCA and to 13/27% in the exclusively AVM feeding right/left ACA. Rates in the brain supplying left MCA remained the same.

In case 2 of Table 11, the left occipital angioma was partially embolized two months prior to its complete removal. A catheter was placed into the malformation via left ICA, left posterior communicating and left PCA. The catheter had to be left in place and resulted in slight narrowing of the left PCA (Fig. 80). After this procedure, the calculated angioma volume had dropped from 20.8 ccm to 6.7 ccm. The flow rate of main feeding left PCA decreased by 34%. The brain supplying right PCA, which had not been filled on pre-embolization vertebral angiography, was excellently visualized on postemboliza-tion control. Its calculated flow rate had increased by 80% (Fig. 80), whereas changes in sonographic flow velocities of both PCAs were negligible. The increase in right PCA flow rate thus was mainly due to vessel enlargement.

The partially AVM supplying left MCA showed a 15% increase of its flow rate following the embolization and was also better visualized on control angiography. Its diameter had increased by 0.04 cm (Fig. 81). Flow velocities in this vessel dropped and its resistance index had increased as a result of the reduced angioma volume. Brain supplying right MCA showed a 39% rise of its flow rate following the embolization.

After complete surgical removal, left PCA flow rate was 11% its initial value, whereas in the right PCA it increased by 93%.

c) Follow-up studies after angioma removal

Statistical evaluation was impossible because of varying size, localization, vessel diameter and hemodynamic characteristics of the different angiomas. We therefore will try to illustrate typical findings in a number of cases with different features. It should be stressed that in all cases the flow velocities in the former AVM feeders dropped dramatically for at least one day after surgery. This was sometimes followed by a short-term and slight acceleration lasting from 3 to 6 days, which again was succeeded by another velocity decrease to normal patterns. Changes in the corresponding contralateral arteries without contribution to the malformation were always slight.

Case 1: Small angioma being supplied by a major MCA branch (Fig. 82)

On the first postoperative day after removal of this small right parietal AVM, the systolic and diastolic flow velocities of the feeding MCA dropped. This corresponds to the angiographically "stagnating" visualization of this vessel. The flow velocity increases on the third day, especially in its systolic component and then almost maintaine its level from the fourth through the eighth day (discharge of the patient). Follow-up studies after 4 weeks and 5 months reveal a slight decrease of right MCA flow velocity, now being almost the

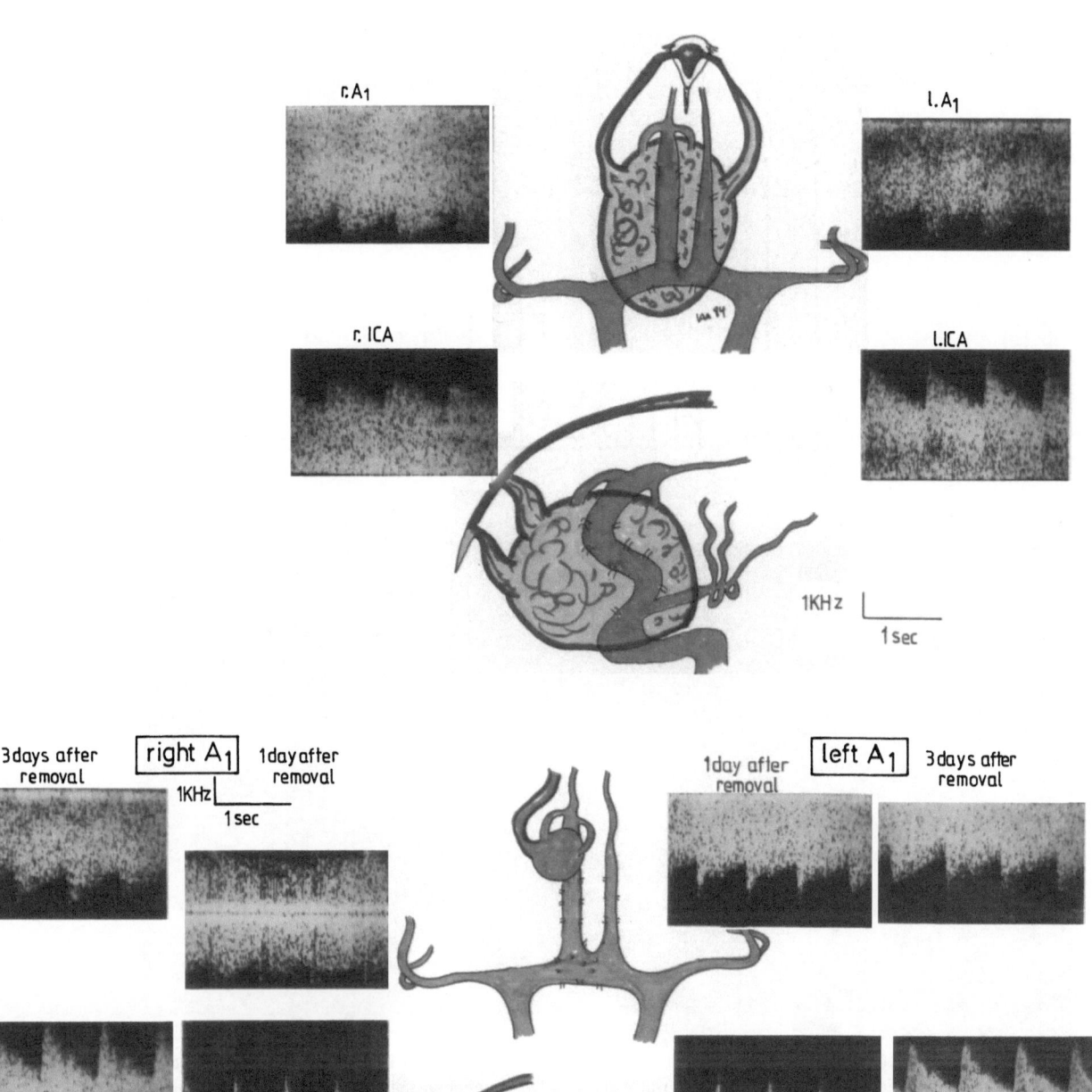

Fig. 79. Flow velocities before and after partial exclusion of a frontobasal medial angioma. Both feeding A1-segments continue to show typical "angioma patterns", although velocities have slightly decreased

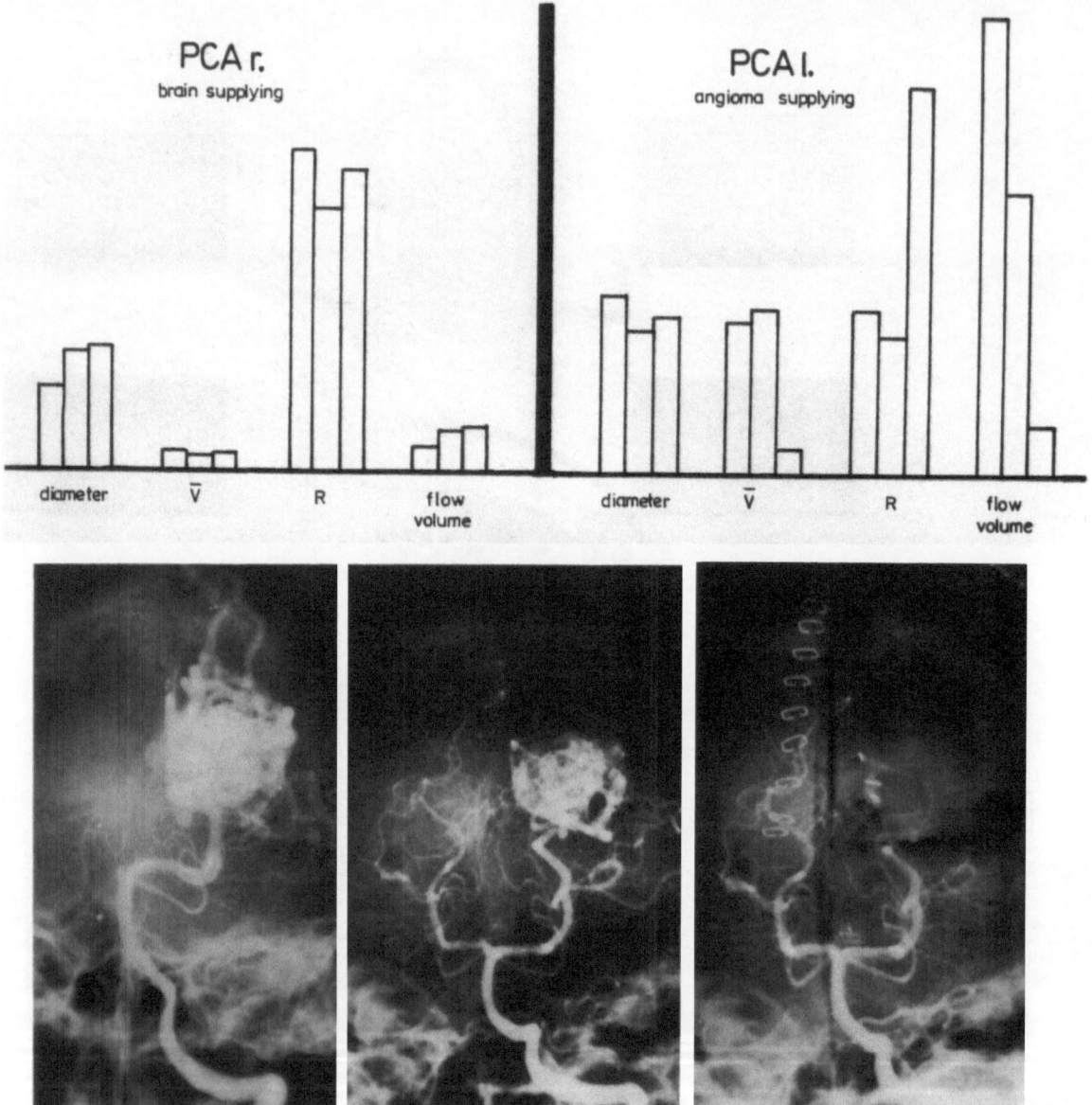

Fig. 80. Diameter, mean flow velocity (v), resistance index (R) and calculated flow volume in right and left PCA before embolization (left), after partial embolization (middle) and after surgical removal (right) of a left occipital angioma; corresponding angiograms

same as in the left MCA. During the whole postoperative course, velocity changes in left MCA were slight.

Angiography was repeated in the fifth postoperative month after an epileptic seizure. Visualization of right MCA was now normal; there were no signs of slowed venous drainage; diameters had normalized.

In this case, adaptation of the right MCA to the new hemodynamic situation took 4 days. From this time on, velocity changes of both MCAs almost ran parallel.

Case 2: Large right Sylvian angioma supplied by right MCA (Fig. 83, same case as Fig. 77)

Immediately following complete surgical removal, the right MCA flow velocity dropped dramatically with enddiastolic zero flow.

The non-AVM supplying left MCA showed signs of increased peripheral resistance with steep systolic peaks and low enddiastolic velocities. This vessel already had a normal flow velocity on the first postoperative day, whereas flow deceleration in the former AVM feeding right MCA persisted until day 16. On long-term follow-up, flow velocities in the affected hemisphere remain below the corresponding contralateral values.

Case 3: Large left occipital angioma supplied by left MCA and left PCA (Fig. 84).

The significance of preoperative Doppler findings for the postoperative course is illustrated by the case of a large occipital angioma that underwent partial embolisation prior to surgery. The left PCA was the main feeding vessel; flow velocities in the left MCA were less

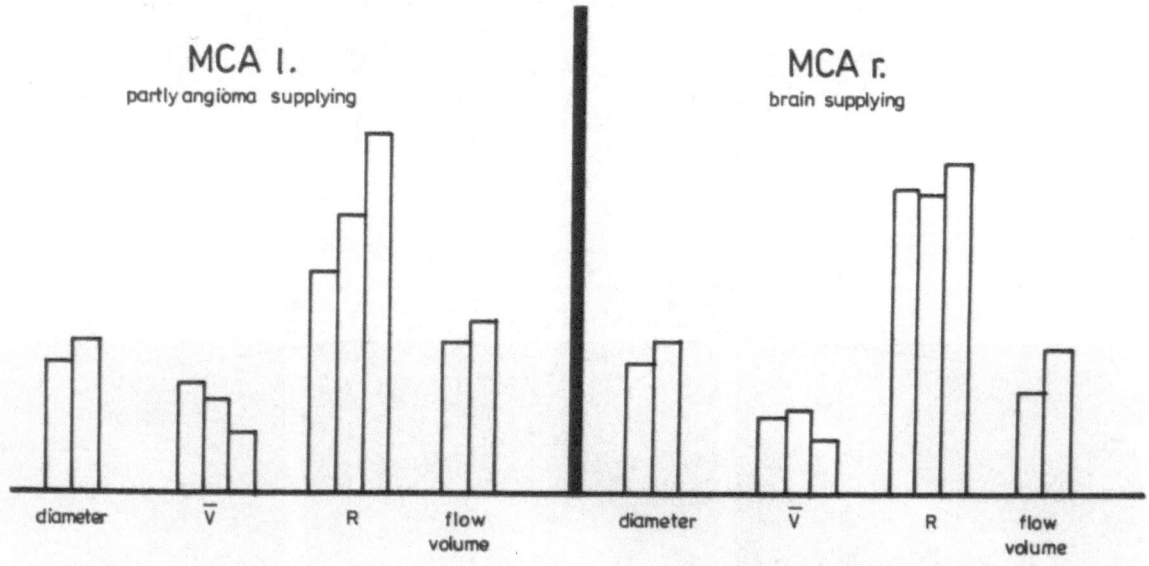

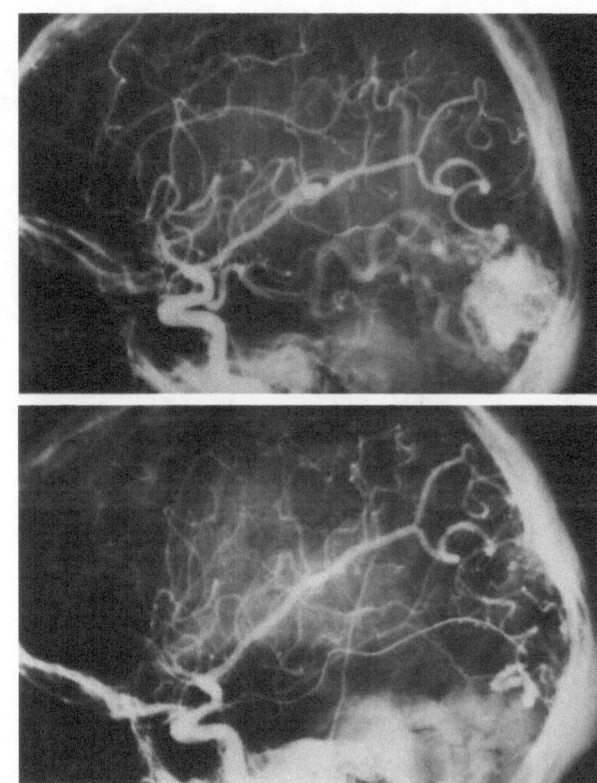

Fig. 81. Diameter, mean flow velocity (v), resistance index (R) and calculated flow volume of right and left MCA before embolization (left), after partial embolization (middle) and after surgical removal of a left occipital angioma (same case as Fig. 79). Note the embolization catheter that had to be left in the left PCA (lower angiogram)

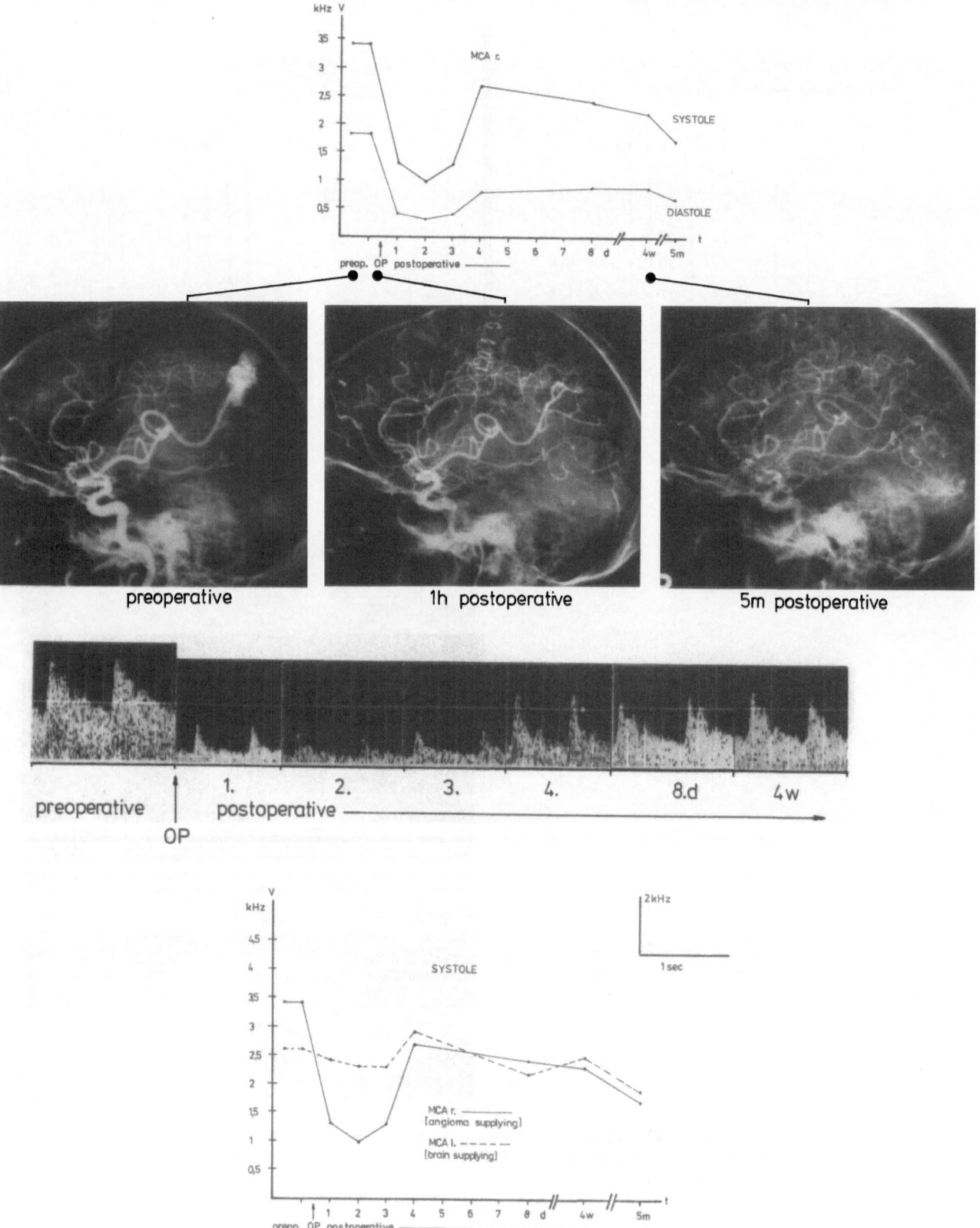

Fig. 82. Flow velocities and angiograms before and after removal of a small right parietal angioma being supplied by the right MCA. Note the typical transitory postoperative flow deceleration in the former feeder (above), whereas the contralateral vessel remains almost unchanged (below)

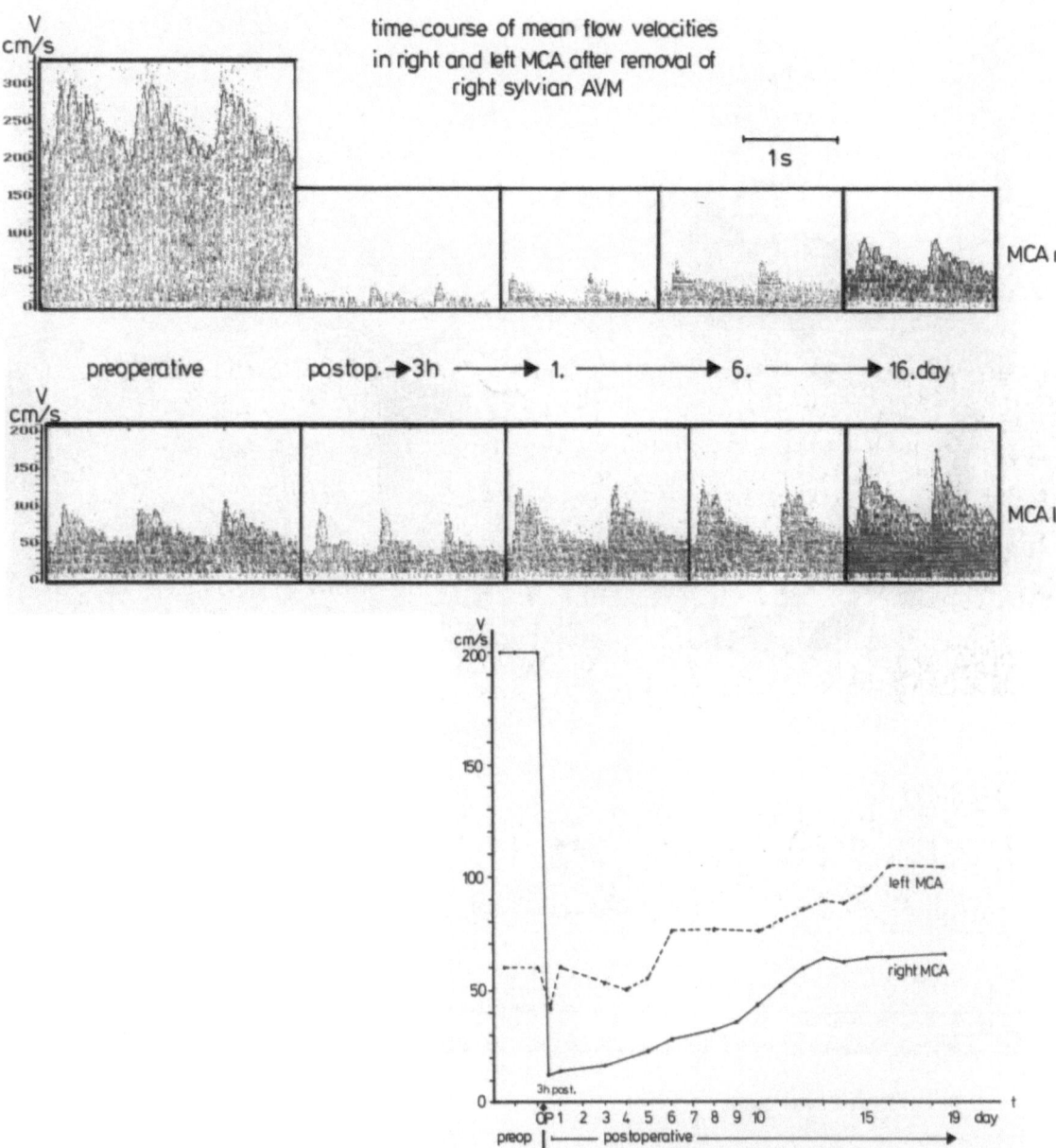

Fig. 83. Course of transitory postoperative flow deceleration in a large right Sylvian angioma (same case as Fig. 77). Three hours after surgery, enddiastolic zero flow is observed in the former feeder. The contralateral vessel also shows signs of increased stream resistance (see text)

elevated. Following AVM removal, the left PCA showed a noticeable drop in flow velocity, followed by the typical slight increase and then normalization on the second postoperative day. Flow velocities in the partially AVM supplying left MCA decreased less after operation and reached normal values within only 24 hours.

The brain supplying right PCA that did not contribute to the malformation started to show rising systolic velocities 3 hours after surgery. Diastolic values remained unchanged. The same effect was seen in the right MCA, where systolic flow acceleration occured from the first through the sixth postoperative day.

Fig. 85 shows the postoperative course of diastolic flow velocities in different cerebral arteries of Case 3, where the left PCA is purely

AVM supplying, the right PCA and right MCA are purely brain supplying and the left MCA is a partially AVM supplying vessel.

d) Changes of peripheral vascular resistance

The resistance index of a cerebral artery is a measure of its peripheral vascular resistance. It is calculated using Pourcelot's formula. We determined the resistance index of the MCA in 70 healthy subjects. The normal value was found to be 0.53–0.57. Changes in peripheral

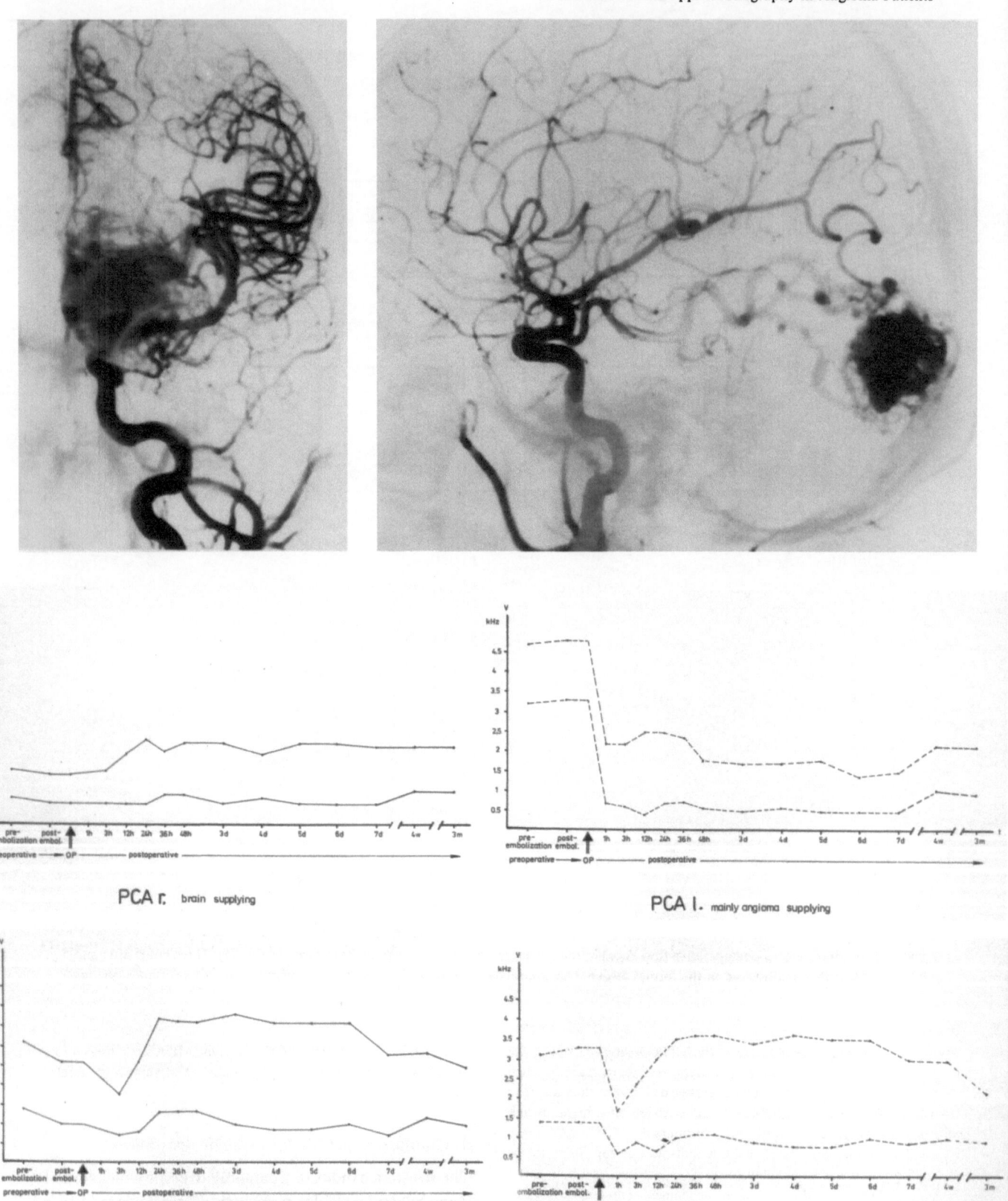

Fig. 84. Angiography and flow velocity changes before and after removal of a large left occipital angioma that had been supplied by the left MCA and left PCA (upper line: systolic velocity, lower line: diastolic velocity)

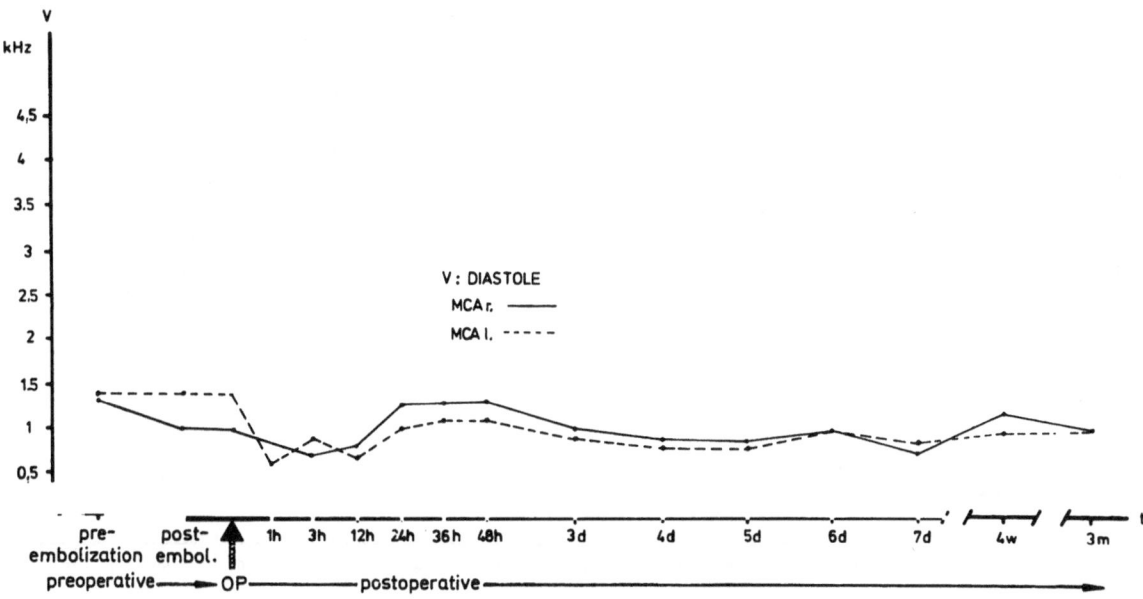

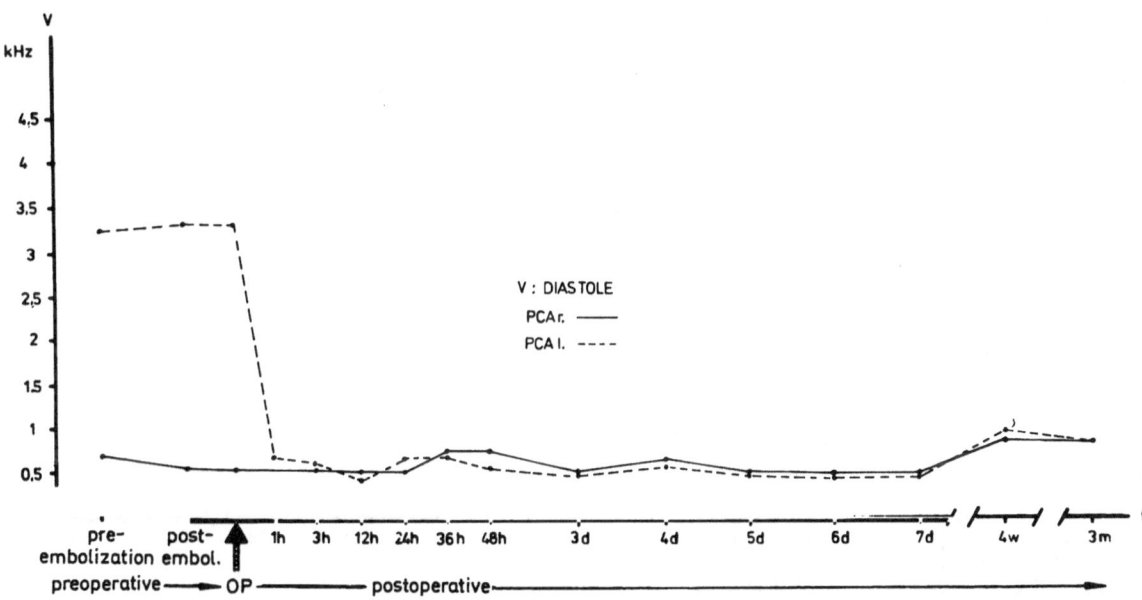

Fig. 85. Comparison of diastolic velocity changes in the same case as Fig. 84: The main feeding left PCA shows a more pronounced flow deceleration when compared to the partially feeding left MCA

vascular resistance after angioma operations depend on the hemodynamic significance of the malformation. Increasing intravascular pressure may be of importance. Some cases are given in the following:

Case 1: Large occipital angioma (Fig. 86)

Postoperative changes in peripheral vascular resistance were most pronounced in the main feeding left PCA, where the resistance index increased from preoperative 0.3 to 0.68 after 1 hour, 0.72 after 3 hours and 0.84 after 12 hours. It slowly declined in the following period, but was still elevated to 0.67 on the seventeenth postoperative day (discharge of the patient). After 4 weeks, the index had normalized (0.52). The partially AVM feeding left MCA shows similar changes, however not as pronounced as in the main feeder.

The contralateral vessels also demonstrated an increase in peripheral stream resistance, which occurred later and was less distinct than on the operated side. Maximal values were found around the seventh day. After 4 weeks. the findings were normal.

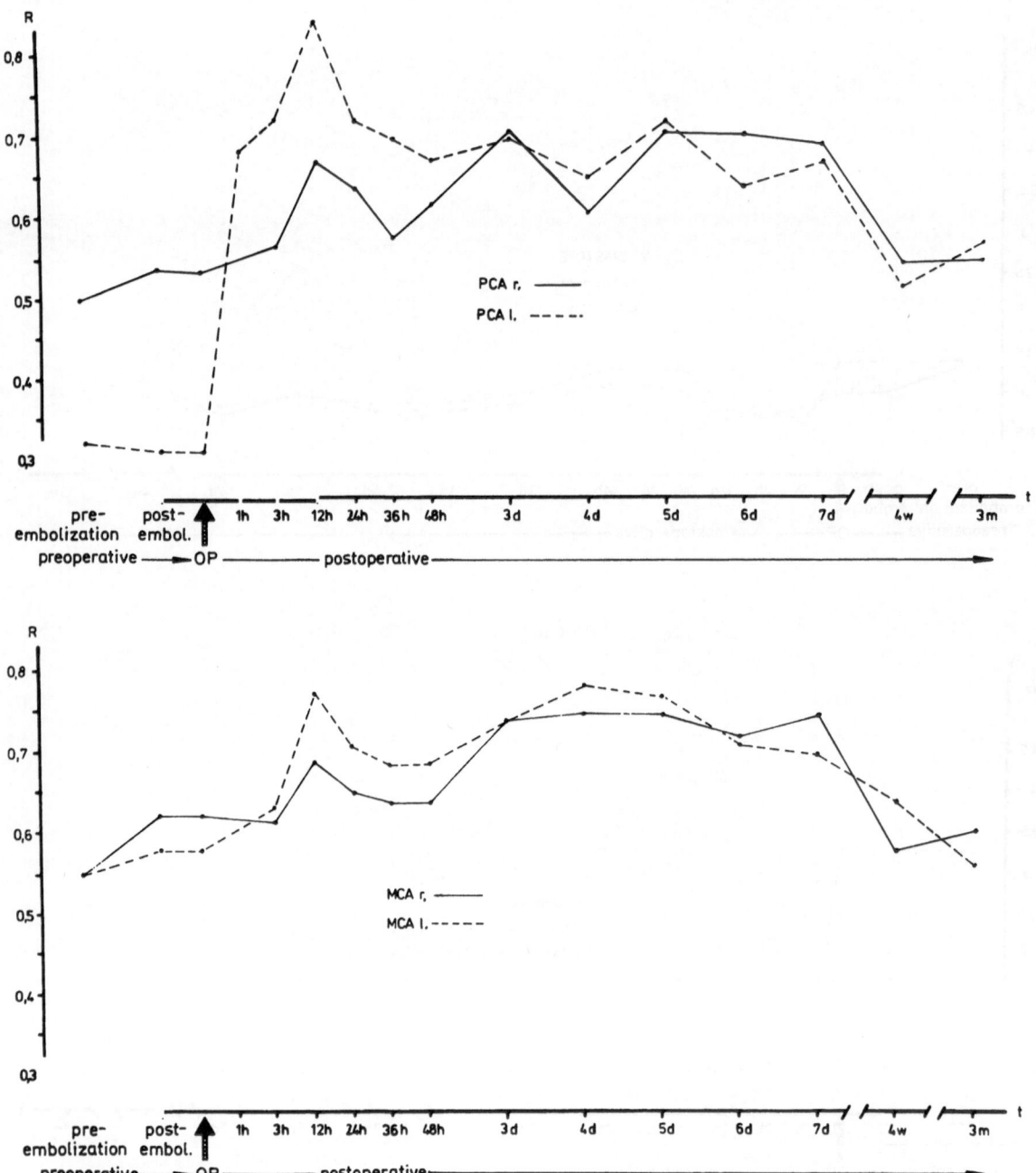

Fig. 86. Resistance indices (*R*) before and after removal of a large left occipital angioma (case 1). The left PCA had been the main feeder, the left MCA had been partially AVM supplying

Case 2: Large frontobasal medial angioma (Fig. 87)

This AVM was supplied by the right MCA and right ACA, which both showed very fast flow. During operation, a hemorrhage into the ventricular and subarachnoid space had occurred.

Three hours after complete AVM removal, the resistance index of the right MCA had increased from 0.39 to 0.85 (Fig. 87). It slowly declined in the following period, but was still found to be elevated on the seventeenth day (discharge of the patient).

The main feeding MCA branch could also be recorded in this patient. On immediate angiography it showed the typical features of a

stagnating artery; the corresponding resistance index was 0.92 at that time. In the repeated angiogram, this stagnating vessel was still visible on the ninth postoperative day when the resistance index was 0.70. As in the previous case, the peripheral stream resistance also increased in the corresponding contralateral vessel reaching 0.64 three hours after surgery and a "delayed" maximum on the fifth postoperative day.

Case 3: Medium-sized parietal angioma (Fig. 88)

This angioma was supplied by the right ACA, right PCA and right MCA. Preoperative flow velocities were moderately increased.

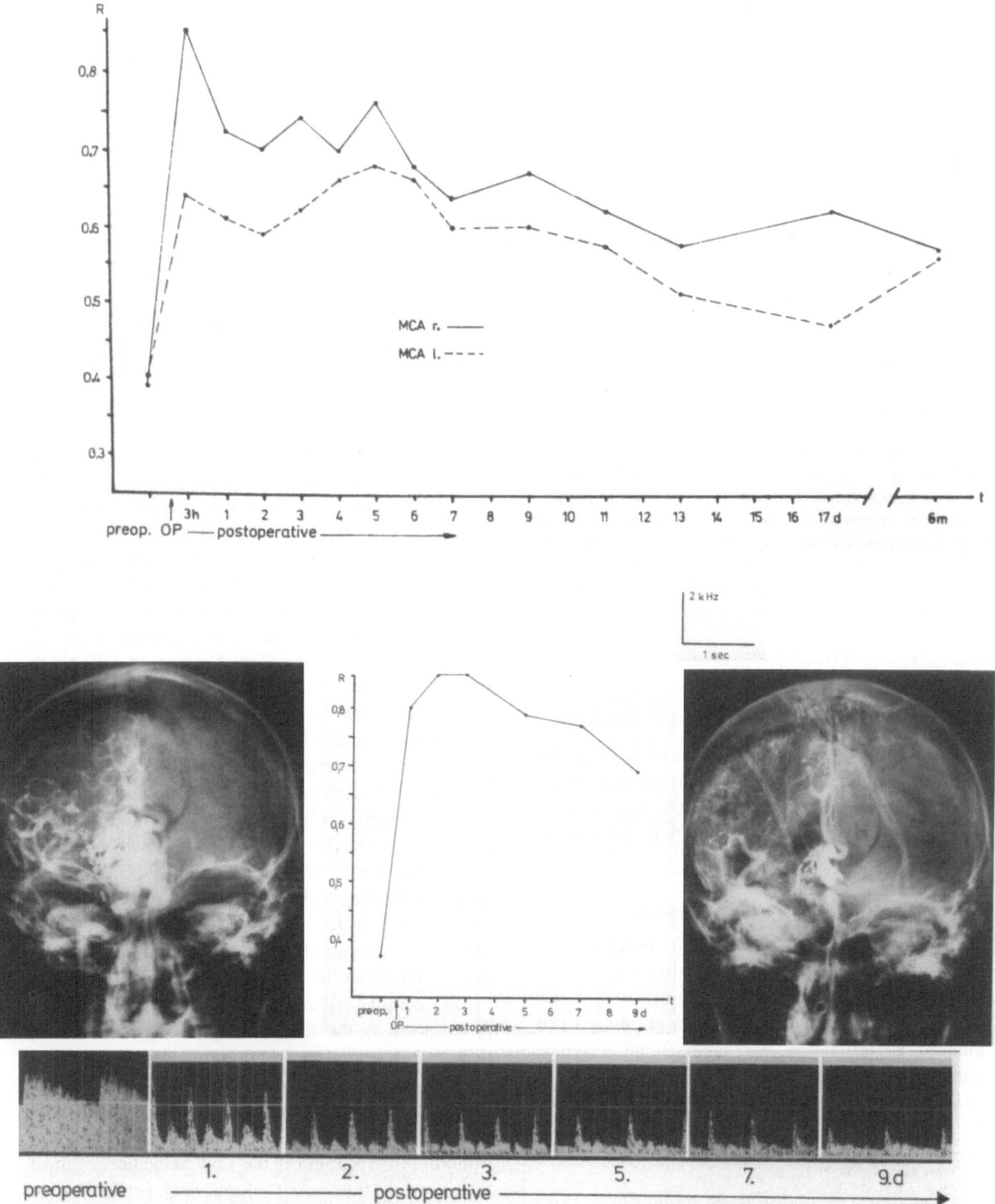

Fig. 87. Above: changes of the resistance index (R) after removal of a frontobasal medial angioma. (Right MCA: AVM feeding, left MCA: brain supplying). Below: typical findings in a stagnating artery (MCA branch)

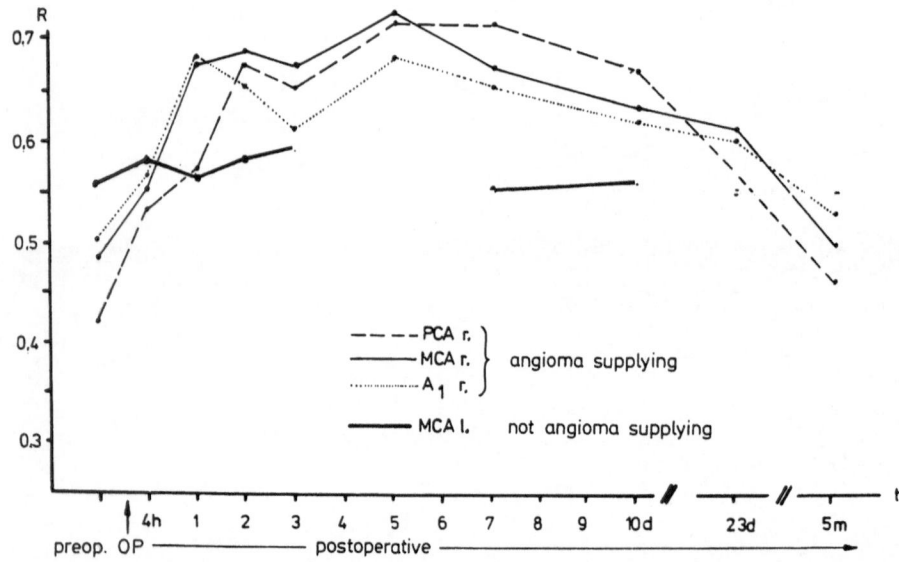

PCA r.　　⎫
MCA r.　　⎬ angioma supplying
A₁ r.　　 ⎭

MCA l.　　not angioma supplying

Fig. 88. Postoperative changes of the resistance indices in the case of a right parietal angioma being supplied by the right MCA, right ACA and right PCA. Changes in the purely brain supplying left MCA are negligible

Postoperative increases in the resistance indices were proportionate to the varying degree of AVM supply of the three feeders. The main feeding right PCA showed the highest postoperative increase from preoperative 0.42 to 0.71 on the fifth day. The right MCA showed a similar course, whereas the changes in the right ACA, which contributed less, were less pronounced. All feeders showed elevated resistance indices throughout the tenth day which normalized within the following 5 months.

2.2.2. Flow Characteristics in Neighboring Brain Arteries After AVM Removal

a) Changes in systolic and diastolic flow velocities

Depending on the size and location of an angioma, postoperative changes in flow velocity occur in neighboring brain arteries. These alterations, however, are slight and always more pronounced in the systolic component of the Doppler spectra.

b) Changes in peripheral stream resistance

The peripheral vascular resistance in the distribution of brain supplying arteries increases after angioma removal. This even applies for the contralateral hemisphere, where the changes, however, are not as pronounced as ipsilateral. We again found a typical biphasic course with initially rising peripheral resistance, then a sharp decrease and a second, smaller rise after 3 days. The reason for this second peak is unknown (Figs. 86–88).

2.2.3. Special Case with Early Hemorrhage After Surgery

A 30-year-old woman was operated on a left parietal angioma. Preoperative transcranial Doppler record-

ings showed high mean flow velocity in the left MCA (112 cm/sec) and normal mean flow velocity in the right MCA (52 cm/sec). Two hours after beginning of the operation, with partial removal of AVM, transcranial Doppler recordings showed decreasing flow velocity in the angioma feeder (MCA left) as well as in the contralateral brain artery (MCA right). With complete removal of the AVM the mean flow velocity of the left MCA dropped far below normal values indicating high stream resistance.

Two hours later, with SAP rising up to 160/100, the left pupill slightly dilated. Transcranial Doppler recordings revealed enddiastolic zero-flow in both MCA. During CT-evaluation, the blood pressure dropped down to 90/30. At this time, transcranial Doppler suggested a circulatory arrest showing oscillating flow in the left MCA whereas the right MCA showed reduced systolic flow. After removal of a large hematoma found on CT, the flow in both MCA was reestablished. The left pupill normalized and the blood pressure was kept at 100/60 in the following days.

Transcranial follow-up recordings showed high flow velocity in the left MCA indicating postischemic hypermia. A second decerebration suddenly occured on the 7th postoperative day. Due to raised ICP the pupills maximally dilated and flow velocities dropped. On both sides only systolic flow was detectable. With induced osmodiuresis (Mannitol) the flow velocities regained their previous (hypermic) values, which persisted up to the 13th postoperative day. In the following period the Doppler patterns stabilized and showed slightly increased resistance indices.

Corresponding CT-scans demonstrate hypodensity

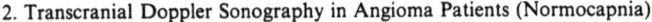

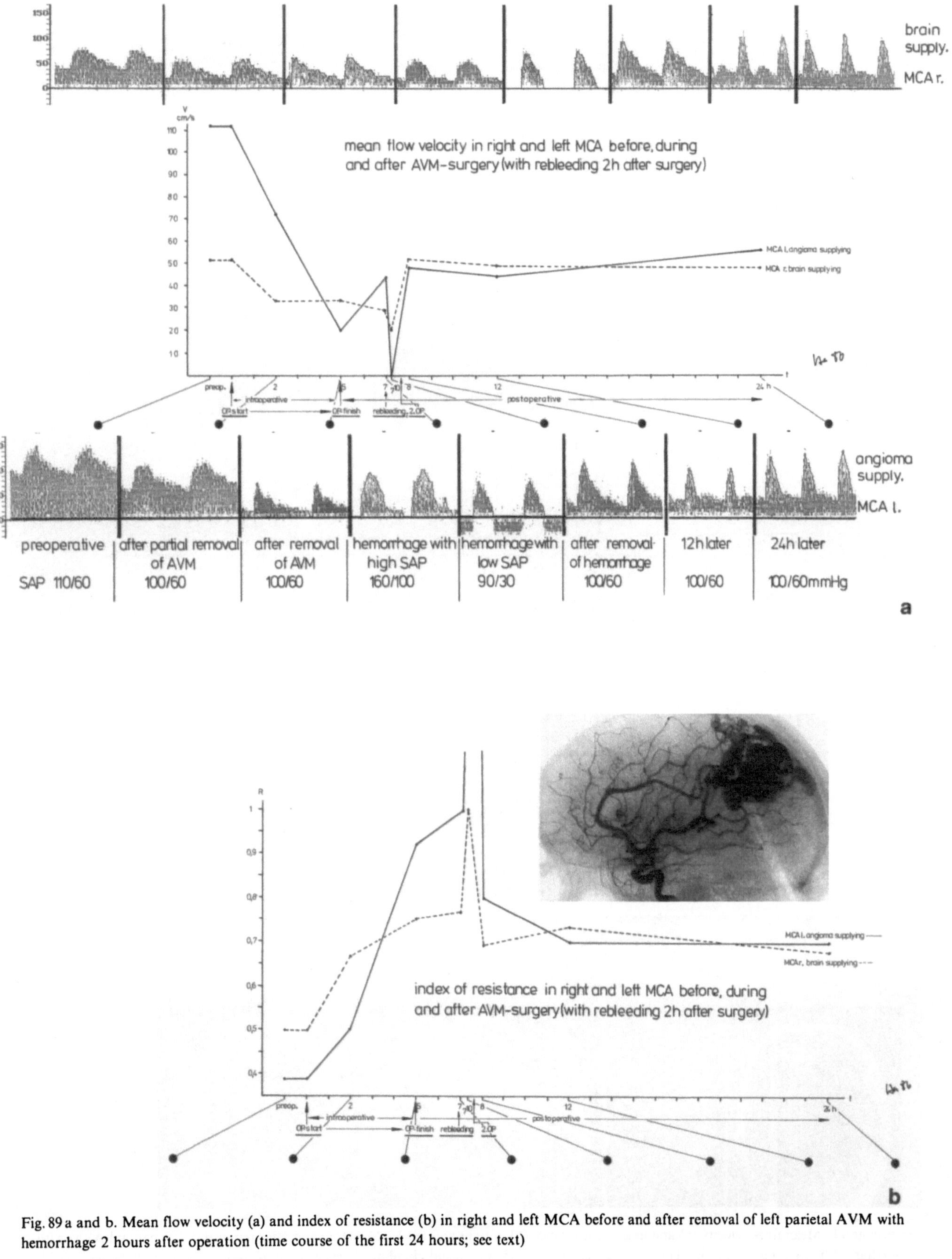

brain supply.

MCA r.

mean flow velocity in right and left MCA before, during and after AVM-surgery (with rebleeding 2h after surgery)

MCA l. angioma supplying

MCA r. brain supplying

V cm/s

preop. 2 6 7 0 8 12 24 h

intraoperative

OP start OP finish rebleeding, 2.OP

postoperative

angioma supply.

MCA l.

preoperative	after partial removal of AVM	after removal of AVM	hemorrhage with high SAP	hemorrhage with low SAP	after removal of hemorrhage	12h later	24h later
SAP 110/60	100/60	100/60	160/100	90/30	100/60	100/60	100/60 mmHg

a

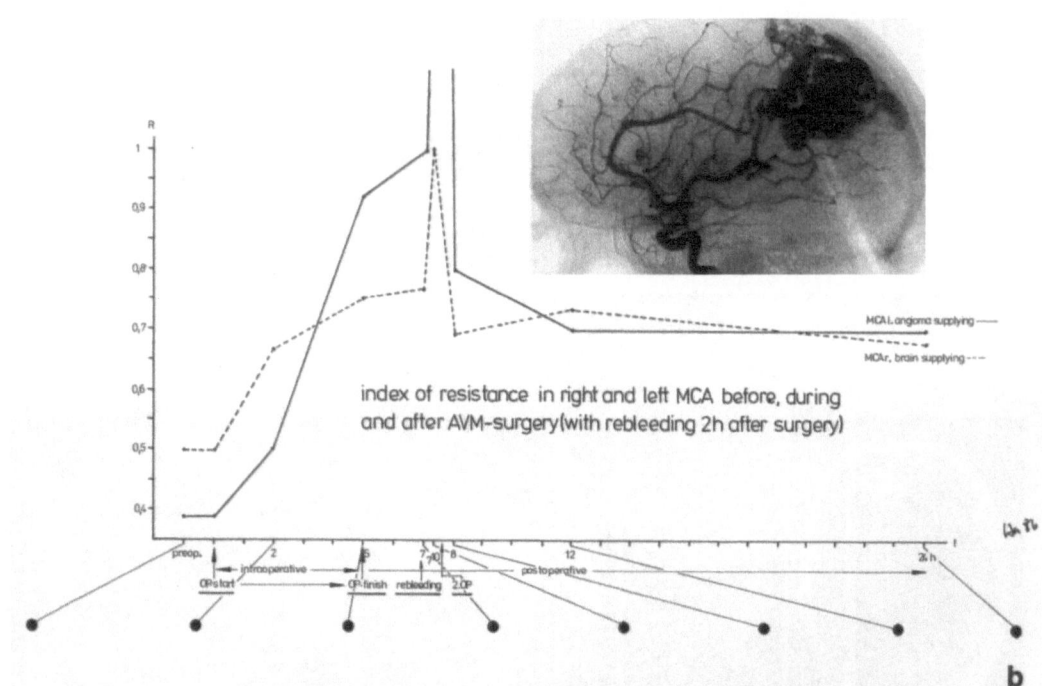

index of resistance in right and left MCA before, during and after AVM-surgery (with rebleeding 2h after surgery)

MCA l. angioma supplying ———

MCA r. brain supplying ‑ ‑ ‑

R

preop. 2 6 7 0 8 12 24 h

intraoperative

OP start OP finish rebleeding 2.OP

postoperative

b

Fig. 89a and b. Mean flow velocity (a) and index of resistance (b) in right and left MCA before and after removal of left parietal AVM with hemorrhage 2 hours after operation (time course of the first 24 hours; see text)

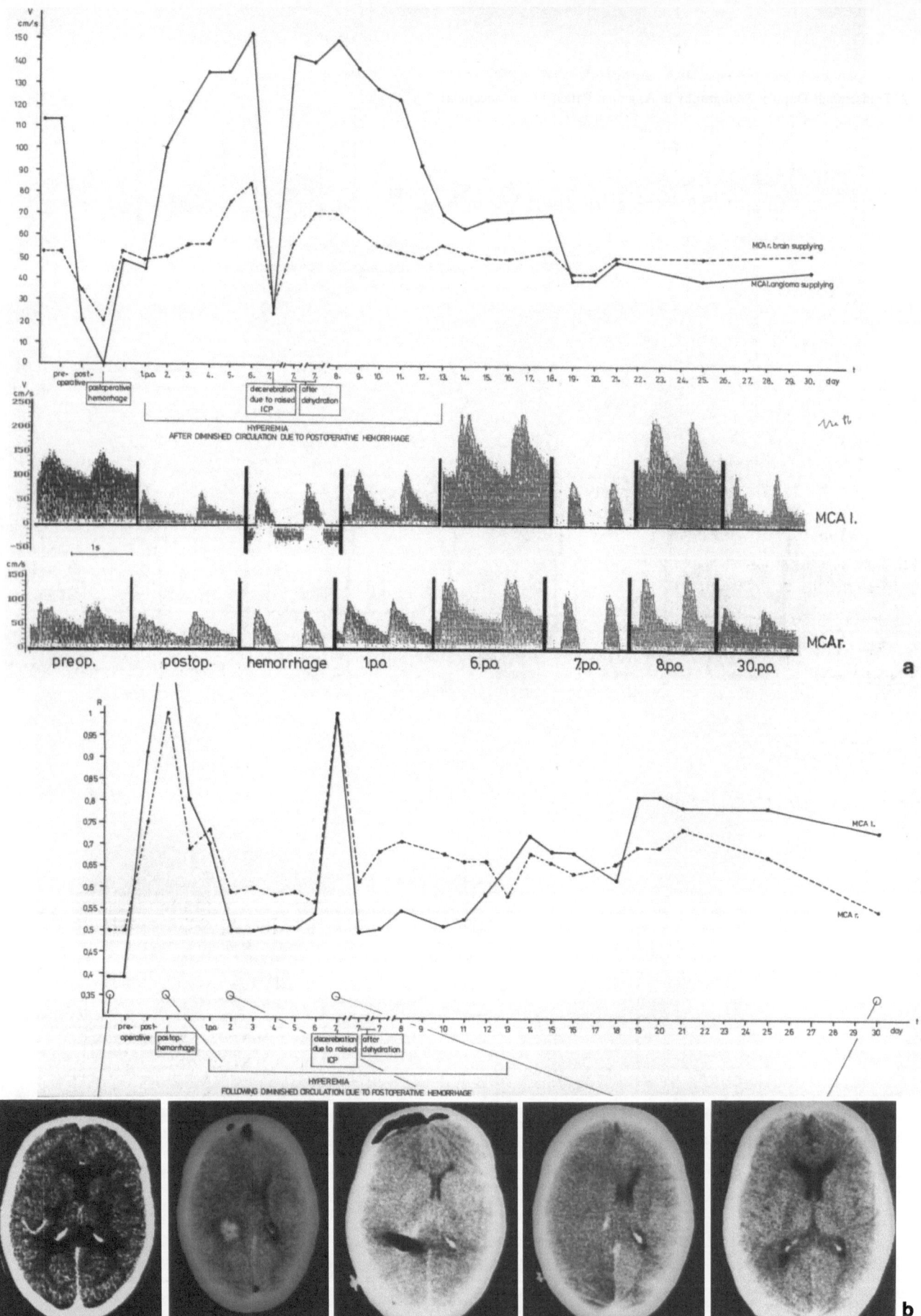

Fig. 90 a and b. Mean flow velocity (a) and index of resistance (b) in right and left MCA before and after removal of left parietal AVM with hemorrhage 2 hours after operation. Time course over 30 days shows partial circulatory arrest (early phase) followed by hyperemia (with decerebration due raised ICP on 7th post-operative day and short term flow deceleration) and recovery.

The corresponding CT-scans show a transitory hypodensity of the left hemisphere which subsides on follow-up and swelling of the whole left hemisphere in the acute phase which subside up to the 30th day. Clinical recovery took much longer than the normalization of the Doppler- and CT-findings.

Angiography at the 4th postoperative day demonstrated the complete AVM-exclusion and patency of all vessels.

This case is the only example in our series showing postoperative hypermia with raised intracranial pressure following AVM removal. However, hyperemia was not caused by vasoparalysis (disturbed autoregulation) secondary to AVM-removal, but followed severe hypoxia due to intracerebral hemorrhage. We feel, that in this case arteriolar vasodilation produced hyperemia, increased cerebral blood volume and thereby raised ICP with consecutive transient decerebration.

2.3. Discussion
(transcranial Doppler sonography, pre- and postoperative evaluations in angioma patients)

Preoperative findings:

There are only two main methods for preoperative hemodynamic evaluation of cerebral angiomas: angiography and transcranial Doppler sonography.

Regional CBF measurements are unsuitable due to poor local resolution. They only give general impressions of changes in brain perfusion (Lassen 1956, Oeconomos 1969, Prosenz 1973, Yamada 1982). CBF studies suggest that a) preoperative perfusion in the AVM surrounding tissue is slightly lower (1%) than in the remaining parts of the ipsilateral hemisphere, b) preoperative CBF in the AVM hemisphere is lower than contralateral (5%), and c) CBF in the hemisphere contralateral to an AVM is lower than normal.

Most commonly, angiomas are classified according to their angiographical characteristics. They are called small with diameters less than 2 cm and large with diameters greater than 4 to 6 cm (Okabe 1983, Guidetti 1980, Wilson 1979, Drake 1979). Luessenhop (1977) was the only author to classify angiomas according to vascular territory and number of feeding arteries.

In our studies, we tried to adopt dynamic criteria for the preoperative assessment of angiomas. We found that feeding arteries can be classified according to their degree of AVM supply:

Purely AVM feeding arteries show very high systolic and diastolic flow velocities as well as low peripheral stream resistance. These features diminish in vessels

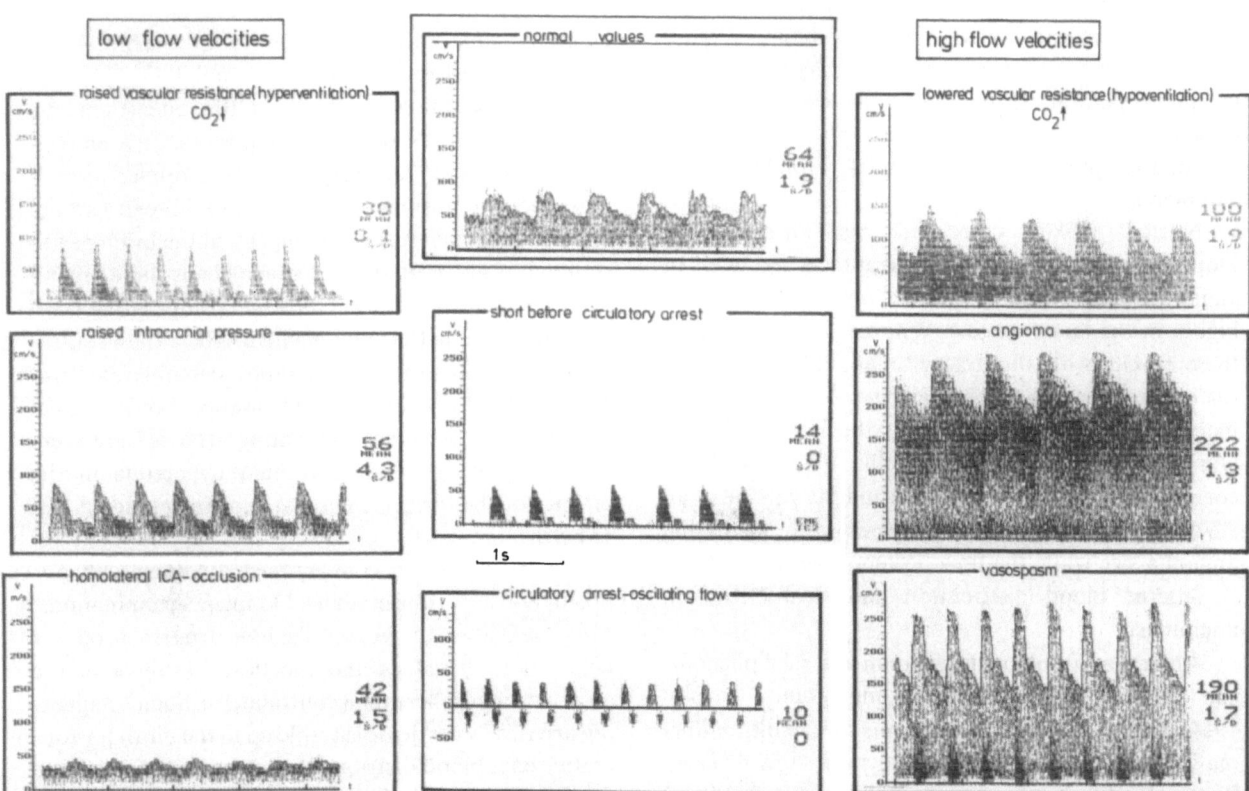

Fig. 91. Transcranial Doppler recordings under various pathological conditions implicating altered intracranial hemodynamics

Table 12: *Transcranial Doppler patterns of arteries with increasing degree of contribution to an AVM: highest flow velocities, lowest resistance indices, and highest diastolic-systolic amplitude ratios are found in pure angioma feeders*

	normal brain supplying	partly angioma supplying	mainly angioma supplying	exclusively angioma supplying	
systolic flow velocities	100	170	180	310	cm/s
diastolic flow velocities	50	110	130	235	cm/s
mean flow velocities	64	132	156	278	cm/s
index of resistance	0,5	0,35	0,27	0,24	
diastolic amplitude / systolic amplitude	50%	65%	72%	76%	

with lesser contribution to an angioma. Feeders with minimal AVM contribution may even equal normal brain arteries.

Another criterion for hemodynamic assessment of angiomas is the flow velocity of brain-supplying arteries. In some of the large angiomas we found diastolic flow acceleration and lowered resistance indices in neighboring "normal" vessels, indicating a compensatory dilation of resistance vessels in brain tissue surrounding the malformation. Systolic velocities were normal in these cases.

Angiography and transcranial Doppler measurements:

Nornes (1980) correlated his intraoperative Doppler findings with the angiographical diameters of angioma feeders and found the flow velocity to be higher in the larger vessels. We were able to confirm these results with our transcranial findings. Nornes further assumed the flow velocity to decrease with increasing length of a feeder due to the frictional forces. Our findings again confirmed this. There is also a correlation between the flow rate and the volume of an AVM. Stream resistance decreases with increasing angioma size while the flow becomes faster.

Altered blood distribution and steal effects in angiomas:

Alterations of blood distribution and steal phenomena are well known from the angiogram (Murphy 1954). Due to low intravascular pressure in the malformation, blood is diverted and may even show reversed flow in some vessels (steal). Such reversed flow is characterized by damped frequency spectra. Com-

pression tests may demonstrate patency of communicating arteries and also the hemodynamic "force" of an angioma. Feeders of large cerebral AVM show only slight Doppler changes with compression of the ipsilateral CCA.

Postoperative findings:

Hemodynamic studies after removal of cerebral AVM have been performed with extracranial Doppler sonography (Büdingen 1977), CBF measurements (Menon 1979, Yamada 1982) and postoperative angiography (Norlen 1979, Hassler 1983). Doppler recordings showed normal flow velocities in extracranial arteries. CBF measurements suggested the brain perfusion to be improved in the angioma hemisphere following the malformation's removal. Postoperative CBF rates are lower in the angioma hemisphere than contralateral. Postoperative angiography demonstrated improved blood distribution and disappearance of steal phenomena. So-called "stagnating arteries" are seen. Some angiograms indicated mild hyperemia in the vicinity of the former angioma, showing enlarged pial vessels.

We were able to confirm such findings with our postoperative transcranial Doppler examinations. Flow velocities in former feeding arteries drop far below normal values and the flow is slower in the angioma hemisphere than contralateral. Such "stagnating arteries" were demonstrable up to the ninth postoperative day; blood distribution was normal. Calculated flow rates of former feeders dropped by 89% compared to the preoperative values.

2.4. Summary

Our transcranial Doppler studies comprise 70 normal subjects and 50 angioma patients. The following results were obtained:

1. Systolic and diastolic flow velocities decrease proportionately with age. Velocities decrease by 32% (systolic) and 33% (diastolic) when the groups of 6–10 years and 60–70 years of age are compared.

2. The vascular resistance index according to Pourcelot does not change with age.

3. Medically induced hypertension in healthy young persons leads to increasing cerebral vascular resistance and cerebral blood flow velocities.

4. AVM-supplying arteries can be distinguished from normal brain arteries.

5. Abnormal findings in angioma feeders depend on their degree of AVM supply: flow velocities are more or less elevated, resistance indices are low and diastolic-systolic velocity ratios high. Feeders can be characterized as vessels of pure, main, partial or little contribution to an angioma.

6. The fastest flow velocities measured in angioma feeders were 6.5 kHz (253.5 cm/sec).

7. Flow velocities in angioma feeders can increase with decreasing recording depth because feeding branches may leave the vessel's main trunk.

8. Turbulent Doppler patterns as well as bruits may be recorded in the vicinity of an angioma and in the malformation itself.

9. Altered cerebral blood flow distribution and steal phenomena are demonstrable in angioma patients. A malformation's hemodynamic impact on normal brain supply becomes assessable with the transcranial Doppler method.

10. Blood flow rates of single vessels can be calculated when flow velocities and angiographic data are known:

—Angioma feeders have approximately ten-fold flow rates in comparison to normal contralateral vessels.
—Flow rates increase with angioma size.
—Flow velocities in feeding arteries increase with angioma size.
—Peripheral stream resistance of the feeders drops with increasing angioma size.
—Flow velocities of AVM feeders decrease with the length of the vessels.
—Peripheral stream resistance of angioma feeders rises with increasing length of the vessels.
—Flow velocities of feeders increase with the vessel diameter.

11. Angioma feeders can develop vasospasm following subarachnoid hemorrhage. Flow velocities further increase under that condition.

12. Angiomas lack vascular autoregulation. Flow velocities change with blood pressure.

13. Pulsatility of the angioma Doppler patterns increases with rising intracranial pressure.

14. Angioma veins show flow velocities up to 39 em/sec. Venous pulsatility diminishes with distance from the malformation.

15. Brain supplying arteries in the surroundings of an angioma may show elevated diastolic flow velocities and slightly decreased resistance indices even though they do not contribute to the malformation.

16. Angioma exclusion can be documented by transcranial Doppler sonography.

17. Partial exclusion of an angioma reduces flow velocities in feeding vessels.

18. After angioma removal, flow velocities in former feeders temporarily drop far below normal values. At the same time, the resistance indices of these vessels are noticeably higher than in normal vessels. These "stagnating" patterns subside in the following period.

19. Postoperatively "stagnating" arteries (former AVM feeders) can be detected with transcranial Doppler sonography up to the ninth day after surgery.

20. The duration of postoperative flow deceleration in former feeders depends on their preoperative degree of AVM supply and on the previous angioma size.

21. Following angioma removal, flow velocities in the ipsilateral hemisphere stay slightly lower than on the contralateral side even on long-term follow-up.

22. Following AVM removal, neighboring brain arteries show slightly reduced or normal Doppler spectra. Signs of hyperperfusion were never seen in these vessels.

3. CO$_2$ Reactivity, Normal Values and Findings in Angioma Patients

Changes in arterial CO$_2$ partial pressure (pCO$_2$) influence the brain perfusion (Kety and Schmidt 1946, Patterson 1955, Lambertsen 1961, Wassermann 1961, Reivich 1964, Harper 1965, James 1969, Grubb 1977, Markwalder 1984). Hypercapnia leads to increased, hypocapnia to decreased perfusion. Angiographically, the diameters of the basal cerebral arteries remain the same under varying pCO$_2$ (Huber 1967). The flow velocity changes measured in these vessels by Doppler sonography must therefore be due to regulations in the peripheral vascular territory where arterioles constrict under hypocapnia (flow deceleration in the basal

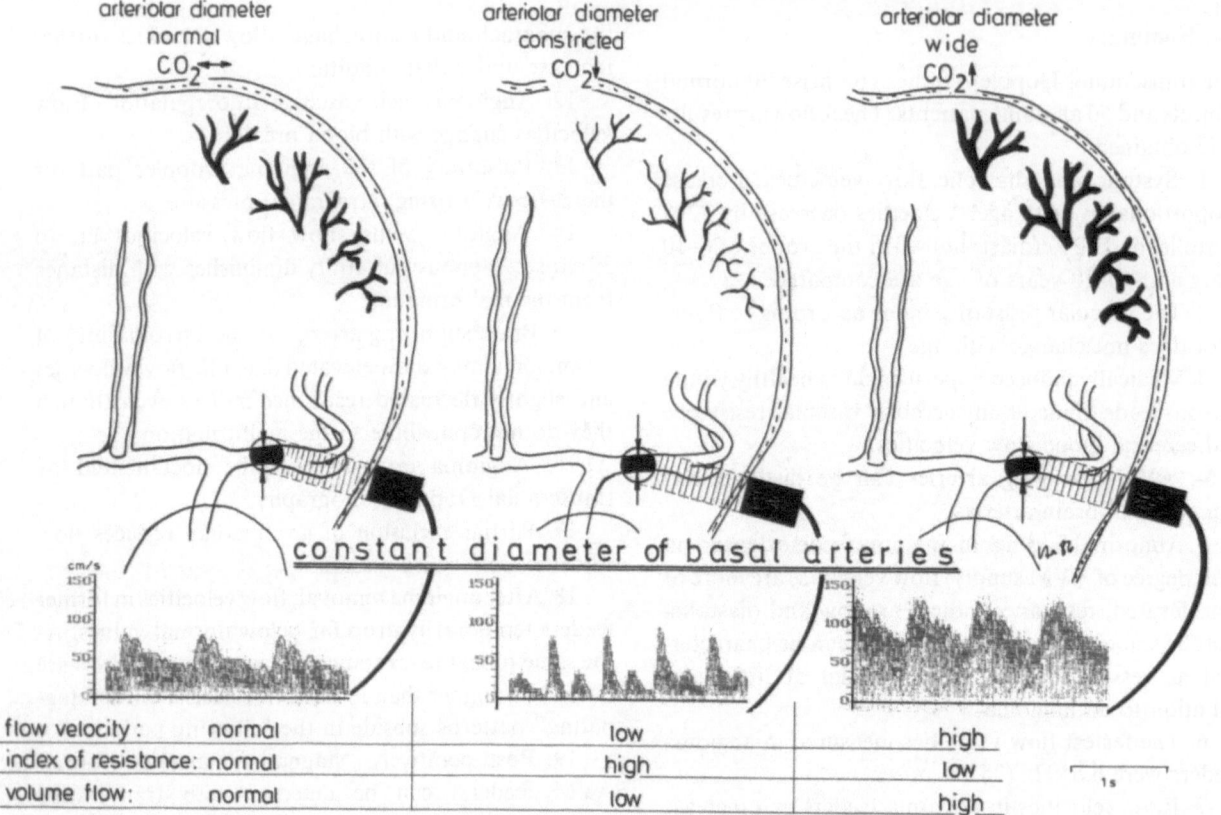

Fig. 92. Flow velocity in MCA in different states of vasomotoric activity. Note: the diameter of basal cerebral arteries remains constant; velocity changes measured in these vessels therefore indicate diameter changes of the peripheral resistance vessels (arterioles)

artery) and dilate under hypercapnia (flow acceleration in the basal artery) (Fig. 92).

Angiomas lack normal arterioles. Intravascular pressure and stream resistance is low in the malformation. As angioma feeding arteries more or less also supply adjacent brain, intravascular pressure should also be low in these areas, leading to compensatory arteriolar dilation and reduced stream resistance in normal brain (Spetzler 1978). Vascular autoregulation in these brain regions is thought to be compromised after long-standing dilatation of the resistance vessels.

We studied the pCO_2-dependent changes in blood flow velocity using the transcranial Doppler technique described above. In angioma patients, arteries supplying both brain and angioma were evaluated before and after surgery. Theoretically, AVM-supplying vessels should have impaired responses to pCO_2 changes, since they lack resistance vessels in their peripheral distribution. By testing the CO_2 reactivity, we therefore attempted to establish additional criteria for the identification of AVM-feeders by Doppler sonography and for the assessment of their hemodynamic significance. From the neurosurgical point of view, examinations of

the postoperative CO_2 reactivity were of special interest. The general assumption is that of the normal perfusion pressure breakthrough theory (Spetzler 1978). Vasoparalysis in the surroundings of the former angioma is thought to result in extravasations leading to brain swelling or bleeding. This hypothesis, however, is insufficiently grounded on experimental findings and has never been verified clinically.

3.1. Method

We studied 70 healthy subjects and 50 angioma patients. Normal values were compiled from seven different age groups of 10 subjects each (see Table 13). Children under the age of 6 were not examined. All measurements refer to the MCA in 4.5 cm recording depth with the exception of children, where recordings had to be made at a lesser depth. In each case, we evaluated the systolic and the diastolic flow velocity and also calculated the corresponding resistance indices according to Pourcelot.

The operative procedure for examinations of the CO_2 reactivity is shown in Fig. 93. The patient (prone position) breathes via a Y-shaped mouthpiece, through which the inhaled and exhaled air are separated from one another. Endexpiratory pCO_2 of the exhaled air is measured by an infrared analyzer (Normocap, Datex Instrumentation Corp., Helsinki, Finnland) and registered by a Hellige recorder.

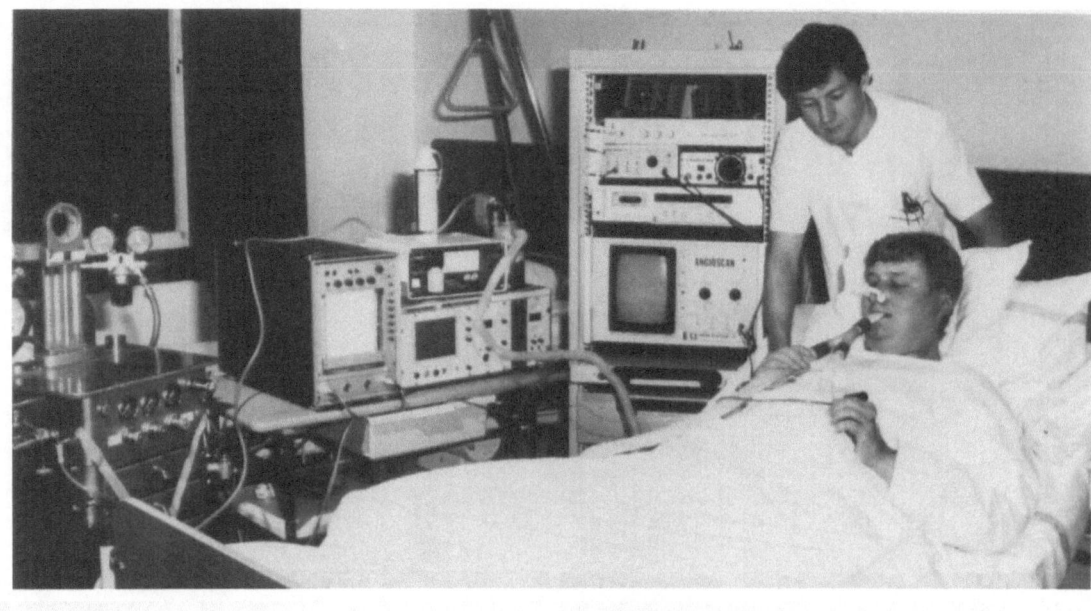

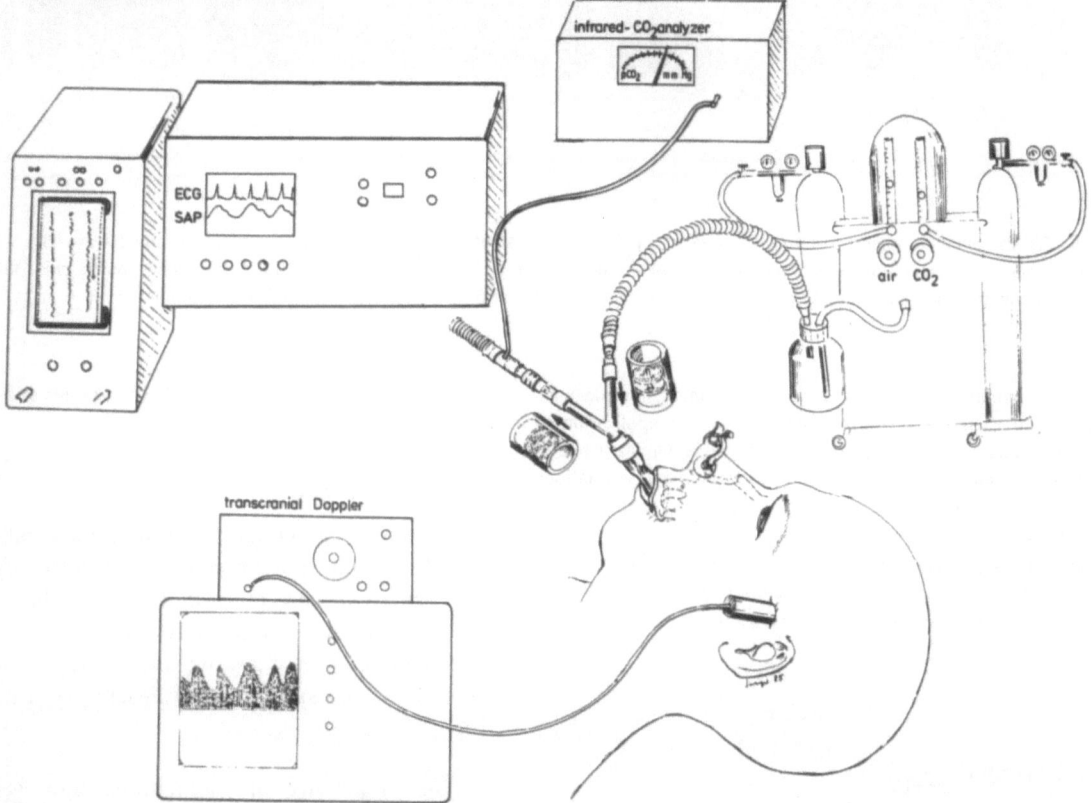

Fig. 93. Equipment for examinations of the CO$_2$ reactivity (see text). I am grateful to Prof. Seeger, Director of the Neurosurgical Clinic of the University of Freiburg for the schematic drawing

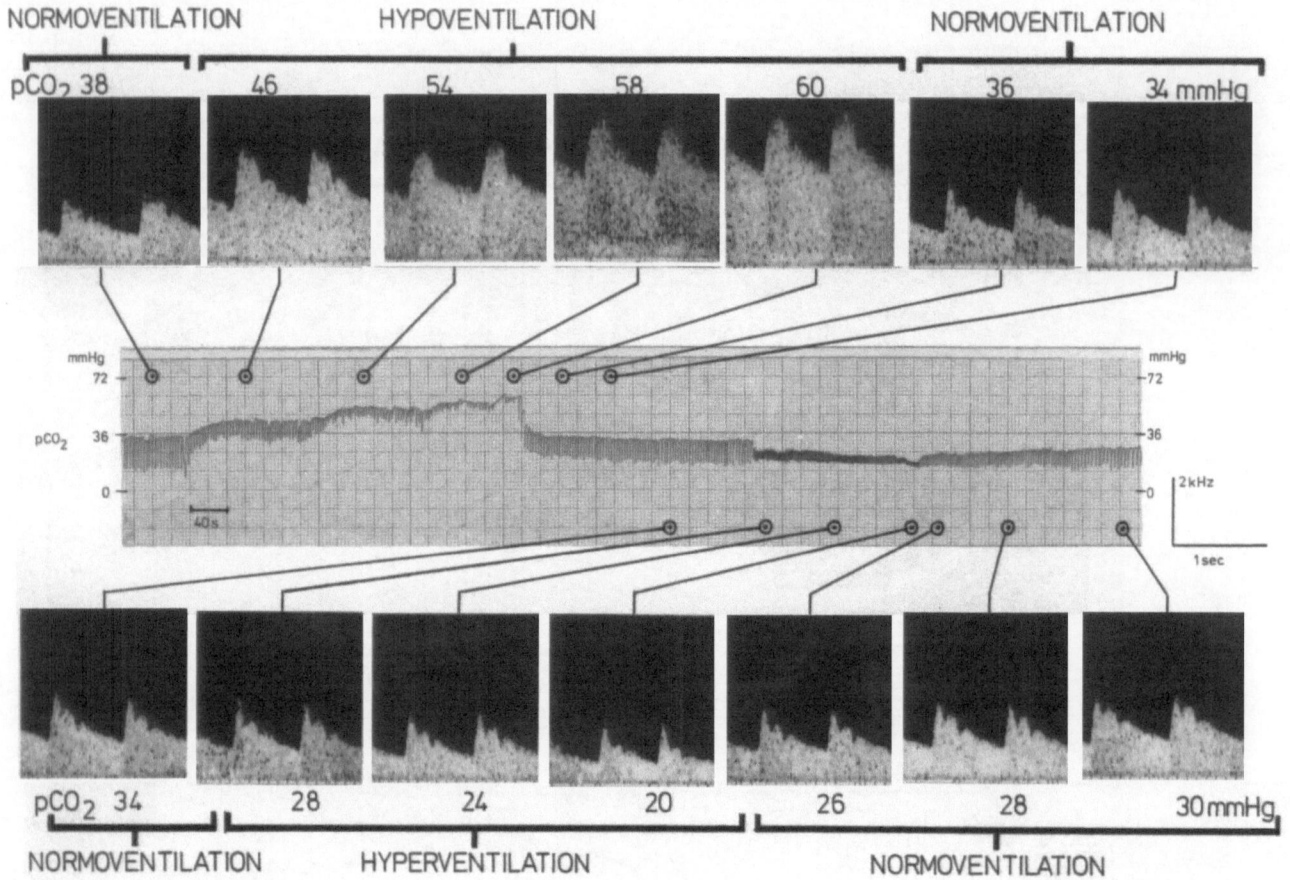

Fig. 94. Transcranial Doppler recordings under varying pCO_2: the examination starts with a normocapnic recording (above left); with induced hypercapnia, continuous Doppler evaluation shows increasing flow velocities, whereas the opposite occurs under hypocapnia. Velocities are always documented after a 2-minute pCO_2 steady state. Endexpiratory pCO_2 values are displayed in the middle

After a 5-minute pause, the recording starts under normocapnia ($pCO_2 = 40 \pm 2$ mmHg). Hypercapnia is induced by application of an air-CO_2 mixture; endexpiratory pCO_2 is raised in steps up to above 60 mmHg. Hypocapnia is induced by spontaneous hyperventilation; step-wise reduction of endexpiratory pCO_2 goes down to 19–22 mmHg.

The typical course of such a procedure is shown in Fig. 94. MCA blood flow velocity was measured continuously. Each time 2 minutes after a certain pCO_2 steady-state is reached, the corresponding flow velocities are documented.

3.2. Terminology

Systolic and enddiastolic flow velocities were each registered under normocapnic, hypocapnic (19–22 mmHg) and hypercapnic (above 60 mmHg) conditions. The range between the maximal systolic flow velocity in hypercapnia and the minimal systolic flow velocity in hypocapnia is called the CO_2 reactivity of the systolic flow velocity. The same applies for the diastolic flow velocity. We do not refer to systolic-diastolic mean velocities. The total capacity of the CO_2 reactivity is the range between the minimal diastolic velocity under hypocapnia and the maximal systolic velocity under hypercapnia.

Our values are given in kHz. One kHz corresponds to a velocity of 39 cm/sec.

3.3. Normal Values

The average systolic and diastolic values and standard deviations of normocapnic MCA flow velocity and the CO_2 reactivities are given in Table 13. They are listed according to age groups of 10 subjects each. The age-dependent course of systolic and diastolic velocities as well as the corresponding CO_2 reactivities are shown in Figs. 97, 98, 99.

a) CO_2 reactivity of the systolic flow velocity in different age groups

The normocapnic systolic velocity decreases until the age of 40. It remains relatively constant in subsequent years of life. The systolic CO_2 reactivity shows only minimal change. It stays at values of 2.4–2.5 kHz. Maximal hypercapnic and minimal hypocapnic velocities show little variations with increasing age.

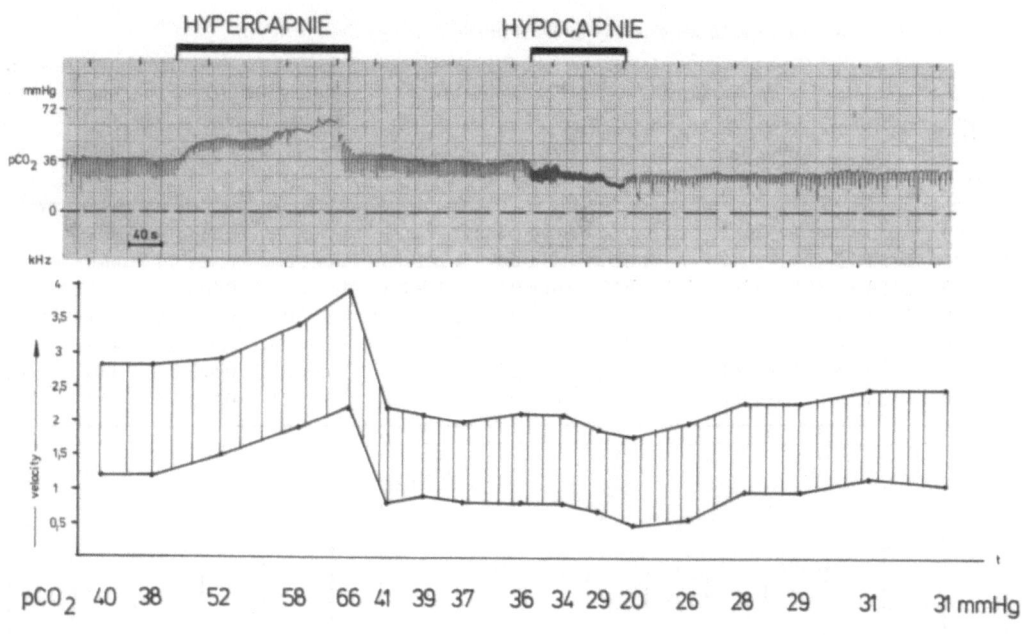

pCO₂ 40 38   52   58 66 41 39 37   36 34 29 20   26   28   29   31   31 mmHg

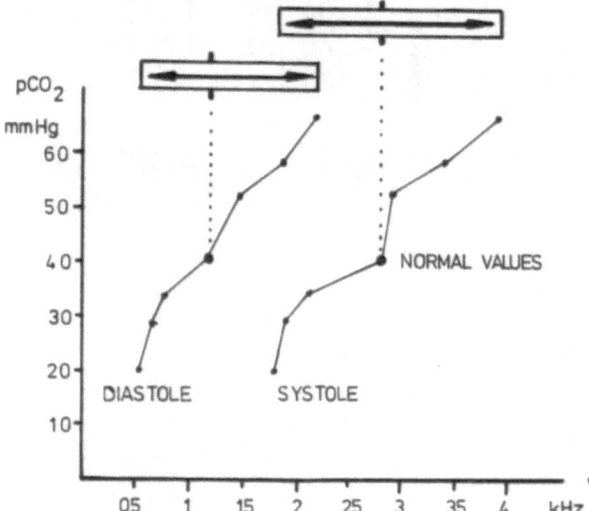

Fig. 95. Changes of systolic and diastolic flow velocities under varying endexpiratory pCO₂ (above). pCO₂-velocity diagram (below)

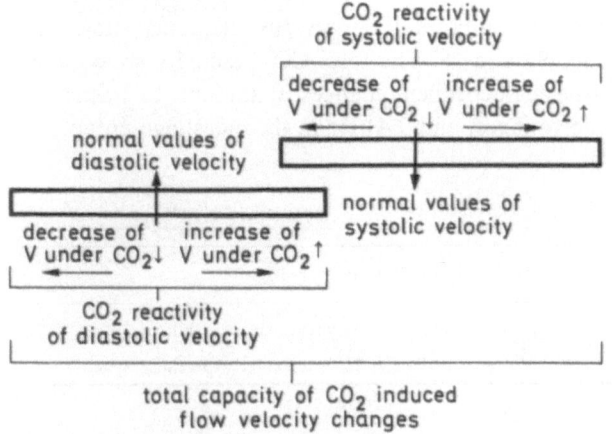

Fig. 96. Terminology (see text)

b) CO₂ reactivity of the diastolic flow velocity in different age groups

Normocapnic diastolic velocities also decrease until the age of 40. Age-dependent diastolic flow deceleration is proportionate to the systolic deceleration; systolic and diastolic normocapnic velocities are reduced by 33% and 32% respectively, when the oldest and the youngest age groups are compared.

The diastolic CO₂ reactivity decreases with age. This is due solely to the reduced ability for flow acceleration under hypercapnia whereas maximal hypocapnic flow deceleration stays the same throughout all age groups reaching minimal values of 0.5 kHz.

Table 13. CO_2 *Regulative (Autoregulative) Capacity of MCA in Different Age Groups (n = 70)*

Age group	n	Mean age (years)	Normal values (kHz)		Hypercapnia (kHz)		Hypocapnia (kHz)		Endexpiratory pCO_2 (mmHg)		
			systole	diastole	systole	diastole	systole	diastole	normal	hyper-capnia	hypo-capnia
6–10	10	7.8 ± 1.5	3.43 ± 0.2	1.5 ± 0.2	4.44 ± 0.3	2.73 ± 0.3	2.08 ± 0.2	0.56 ± 0.1	38.1	63.6	20
11–20	10	16.3 ± 3.3	3.29 ± 0.5	1.49 ± 0.5	1.49 ± 0.3	4.29 ± 0.5	2.49 ± 0.4	0.52 ± 0.1	40.6	68.7	20.3
21–30	10	24.8 ± 3.1	2.76 ± 0.3	1.3 ± 0.2	4.01 ± 0.5	2.29 ± 0.3	1.85 ± 0.3	0.56 ± 0.1	42.4	66.4	21.1
31–40	10	33.1 ± 1.7	2.39 ± 0.2	1.11 ± 0.1	3.74 ± 0.3	2.13 ± 0.3	1.81 ± 0.2	0.6 ± 0.01	41.1	65.7	22
41–50	10	45.6 ± 2.6	2.41 ± 0.4	1.13 ± 0.2	3.82 ± 0.4	2.0 ± 0.2	1.75 ± 0.3	0.57 ± 0.1	41.4	68.4	22.1
51–60	10	55.5 ± 2.5	2.38 ± 0.3	1.1 ± 0.1	3.92 ± 0.2	2.04 ± 0.1	1.75 ± 0.3	0.57 ± 0.1	40.0	64.1	21.1
61–70	10	67.3 ± 5.6	2.33 ± 0.1	1.0 ± 0.2	4.0 ± 0.2	2.03 ± 0.1	1.83 ± 0.1	0.56 ± 0.1	40.3	70	21.3

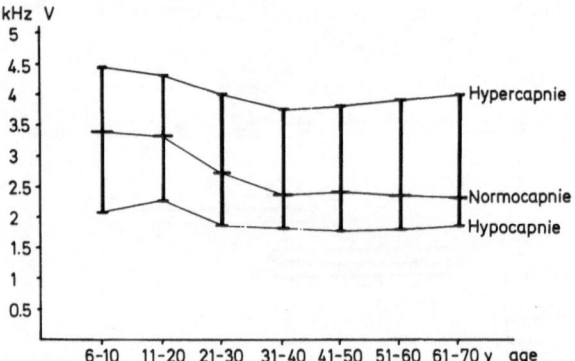

Fig. 97. CO_2 reactivity of the systolic flow velocity in different age groups (MCA)

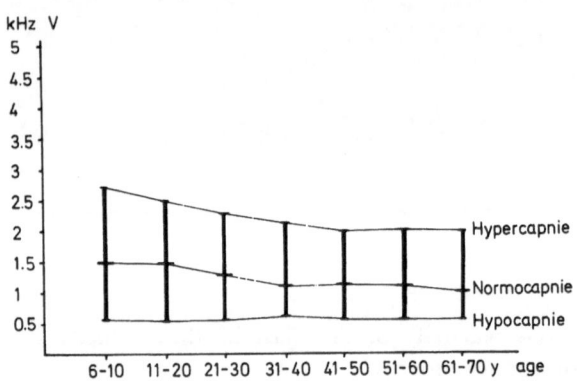

Fig. 98. CO_2 reactivity of the diastolic flow velocity in different age groups (MCA)

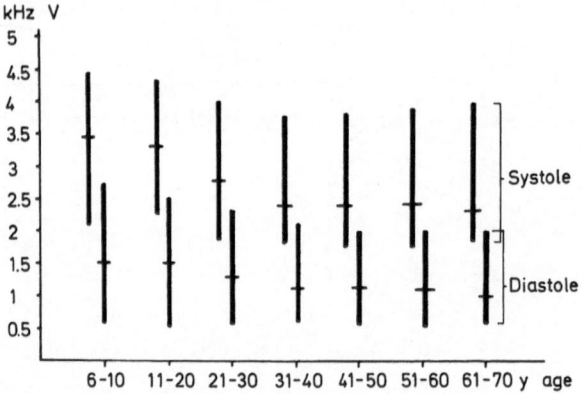

Fig. 99. CO_2 reactivity of systolic and diastolic flow velocities in different age groups. Upper vertical lines: reactivity of the systolic velocity. Upper horizontal lines: normocapnic systolic velocity. Lower vertical lines: reactivity of the diastolic velocity. Lower horizontal lines: normocapnic diastolic velocity

c) Total capacity of CO_2 reactivity in different age groups

Fig. 99 combines the previous diagrams shown in Figs. 97 and 98. The total CO_2 reactivity shows only little age-dependent changes. It amounts to 3.8 kHz in the youngest and 3.4 kHz in the oldest age group.

Table 14. *Changes (%) of MCA Flow Velocity under Hypercapnia in Different Age Groups (n = 20)*

	Age group 6–10 years (n = 10)			Age group 61–70 years (n = 10)		
	normocapnia kHz	hypercapnia kHz	change %	normocapnia kHz	hypercapnia kHz	change %
Systole	3.43	4.44	29.4	2.33	4.0	71.6
Diastole	1.5	2.7	80	1.0	2.03	103

Table 15. *Changes (%) of MCA Flow Velocity Under Hypocapnia in Different Age Groups (n = 20)*

	Age group 6–10 years (n = 10)			Age group 61–70 years (n = 10)		
	normocapnia kHz	hypocapnia kHz	change %	normocapnia kHz	hypocapnia kHz	change %
Systole	3.43	2.08	64.9	2.33	1.83	27.3
Diastole	1.5	0.56	167.8	1.0	0.56	78.5

Table 16. *CO$_2$-regulative Capacity in Different Age Groups (n = 20)*

	Age group 6–10 years (n = 10)		Total regulation	Age group 61–70 years		Total regulation	Age-dependent reduction %
	hyper-capnia	hypo-capnia		hyper-capnia	hypo-capnia		
Capacity diastole	4.44–	2.08	→ 2.36 kHz	4.0–	1.83	→ 2.17 kHz	8
Capacity diastole	2.73–	0.56	→ 2.17 kHz	2.03–	0.56	→ 1.47 kHz	32

Table 17. *Resistance Index of MCA (according to Pourcelot) in Different Age Groups under Normocapnia, Hypercapnia and Hypocapnia (Normal Values)*

Age groups	Normo-capnia	Hyper-capnia	Hypo-capnia
6–10	0.56	0.38	0.73
11–20	0.55	0.42	0.77
21–30	0.53	0.43	0.70
31–40	0.53	0.43	0.70
41–50	0.53	0.48	0.67
51–60	0.54	0.48	0.62
61–70	0.57	0.49	0.69
pCO$_2$ (mm Hg)	40.3	65	21

As shown in Table 17, the normocapnic RI (MCA) remains almost unchanged with age (average value of 0.54–0.55). Under hypocapnia, the peripheral stream resistance rises as a result of arteriolar constriction. The resistance indices therefore increase. Arterioles dilate under hypercapnia, which results in decreasing resistance indices. The age-dependent course of the average maximal (hypercapnic) and minimal (hypocapnic) RI is shown in Fig. 100.

d) pCO$_2$-dependent changes in the resistance index, age-dependent course

The resistance index according to Pourcelot (RI) was calculated from the systolic and enddiastolic flow velocities given in Table 13. It represents a measure of the peripheral vascular resistance in the distribution of the evaluated artery and is mainly influenced by the enddiastolic value.

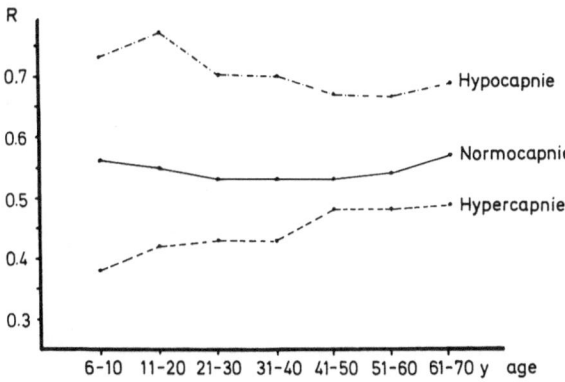

Fig. 100. Resistance index under varying pCO$_2$ in different age groups (MCA)

Table 18. *Changes (%) of Resistance Index (R) in MCA under Hypercapnia and Hypocapnia in Different Age Groups*

Age group 6–10 years (n = 10)			Age group 61–70 years (n = 10)		
Normocapnia	hypercapnia	% difference	normocapnia	hypercapnia	% difference
0.56	0.38	−32%	0.57	0.49	−14%
Normocapnia	hypocapnia	% difference	normocapnia	hypocapnia	% difference
0.56	0.73	+30%	0.57	0.69	+21%
Total change of R in hypercapnia and hypocapnia					
	92%			41%	

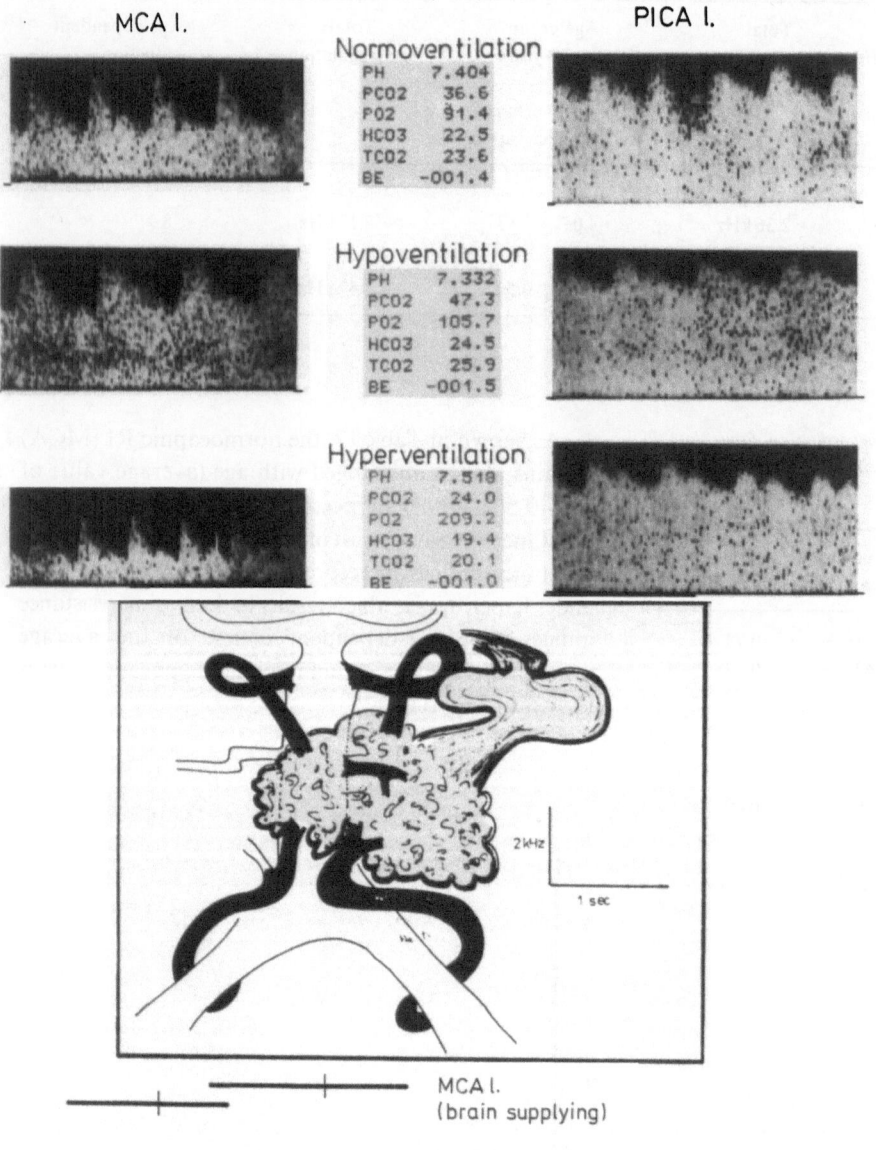

Fig. 101. CO_2 reactivities in the case of an angioma supplied by both PICA: Left PICA shows very poor systolic and diastolic CO_2 responses; normal findings in purely brain supplying left MCA

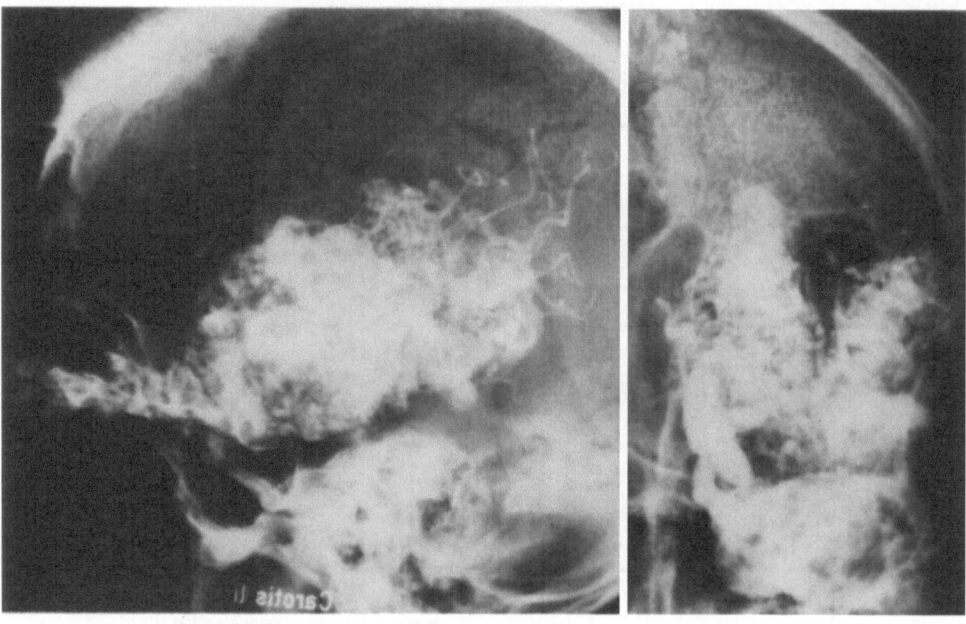

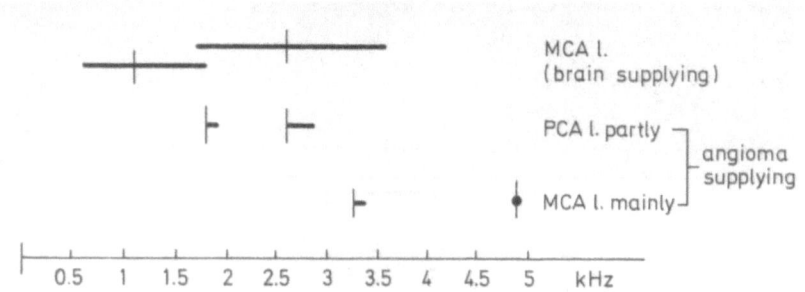

Fig. 102. CO$_2$ reactivities in the case of a left Sylvian AVM. The left MCA is an exclusively AVM feeding artery, left PCA has partial contribution, right MCA is purely brain-supplying. Note the absent diastolic response of the left MCA and the normal findings in right MCA

The ability for hypocapnic arteriolar constriction becomes slightly reduced until the age of 50. More pronounced, however, is the decreasing ability for arteriolar dilation under hypercapnia. Children show a minimal hypercapnic RI of 0.38, whereas this value is 0.48 in the 41–50-year-olds.

3.4. CO$_2$ Reactivity in Angioma Patients

In 50 angioma patients, the CO$_2$ reactivity of basal cerebral arteries was studied. All patients also had cerebral angiography. Transcranial Doppler evaluations were conducted on all basal vessels using the technique described above. In some cases, recordings from more superficial, AVM-feeding branches were also obtained. Hypocapnia or hypercapnia was induced as described in the control group.

In general, angioma-feeding basal cerebral arteries show strikingly increased frequencies upon Doppler evaluation. There are, however, a few cases of usually small AVMs in distal locations with normal Doppler findings in vessels which angiographically supply the angioma. The only dopplersonographical abnormality in such patients may be a side difference of more than 0.5 kHz between the corresponding arteries. The same may in part apply for larger angiomas, where the main feeders are easily identified, but subordinate feeders of mainly brain supply can have unremarkable flow patterns. Side differences of more than 0.5 kHz were considered pathological by Arnold (1986); even they, however, were, absent in some of our cases.

We found the impaired CO$_2$ reactivity of such vessels to be a more sensitive criterion. The following classification has been developed:

a) Exclusively angioma-supplying arteries

CO$_2$ reactivity of such vessels is almost completely abolished (Figs. 101, 102, 104). Diastolic velocities increase slightly under hypercapnia, whereas the systolic velocity is unresponsive. Hypocapnia does not cause any changes. The total capacity of the CO$_2$ response was below 0.5 kHz in these vessels (Table 19).

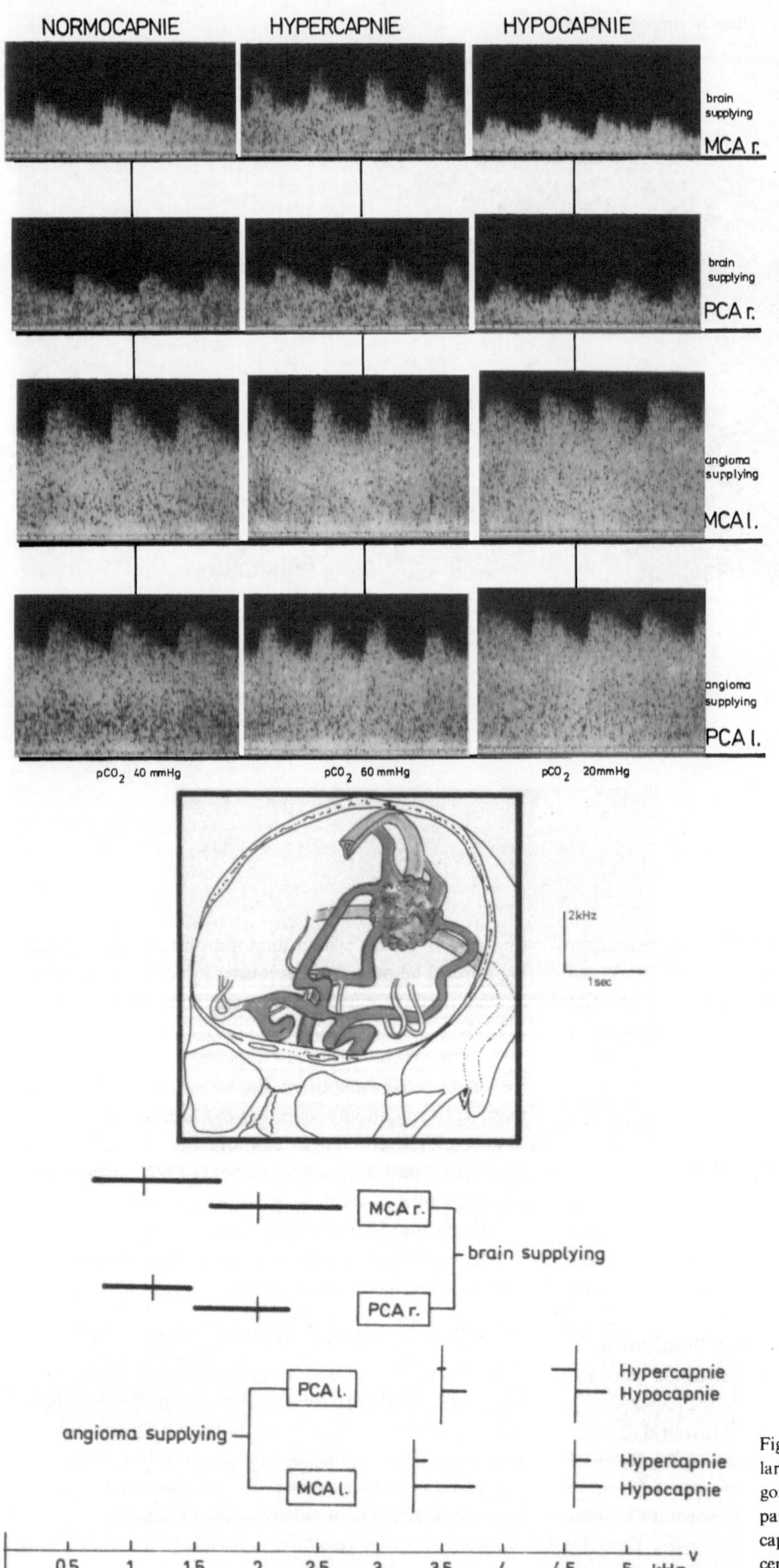

NORMOCAPNIE HYPERCAPNIE HYPOCAPNIE

brain
supplying
MCA r.

brain
supplying
PCA r.

angioma
supplying
MCA l.

angioma
supplying
PCA l.

pCO$_2$ 40 mmHg   pCO$_2$ 60 mmHg   pCO$_2$ 20mmHg

2kHz

1sec

MCA r.
— brain supplying
PCA r.

PCA l.   Hypercapnie
angioma supplying   Hypocapnie

MCA l.   Hypercapnie
                    Hypocapnie

0.5 1 1.5 2 2.5 3 3.5 4 4.5 5 kHz v

Fig. 103. CO$_2$ reactivities in the case of a large angioma of the left ventricular trigonum supplied by left MCA and left PCA: paradoxical flow acceleration under hypocapnia occurs in the AVM feeders (intracerebral steal, see text)

	Normocapnic flow velocity V (kHz)		CO$_2$ reactivity (kHz)		Capacity of CO$_2$ regulation (kHz)
	systole	diastole	systole	diastole	total capacity
Purely angioma supplying	> 4	≥ 3	CO$_2$↑ → very small V↑ CO$_2$↓ → very small V↓	CO$_2$↓ → very small V↑ CO$_2$↓ → V↔	0.5–1 greatly reduced
Mainly angioma supplying	3.5–4	2.5–3	CO$_2$↑ → V↑ CO$_2$↓ → V↓ slightly reduced	CO$_2$↑ → V↑ CO$_2$↓ → V↔ greatly reduced	1–2 greatly reduced
Partially angioma supplying	2.5–3.5	1.7–2.5	CO$_2$↑ → V↑ CO$_2$↓ → V↓ normal	CO$_2$↑ → V↑ (normal) CO$_2$↓ → V↓ (reduced)	2–3 slightly reduced
	normal values	normal values*	CO$_2$↑ → V↑ CO$_2$↓ → V↓	CO$_2$↑ → V↑ (normal) CO$_2$↓ → V↓ (not under	3–4
Minimally angioma supplying	2–3.4 with side differences above 0.5 kHz	1–1.5	normal	I kHz)	normal

* Depending on age. V = flow velocity.

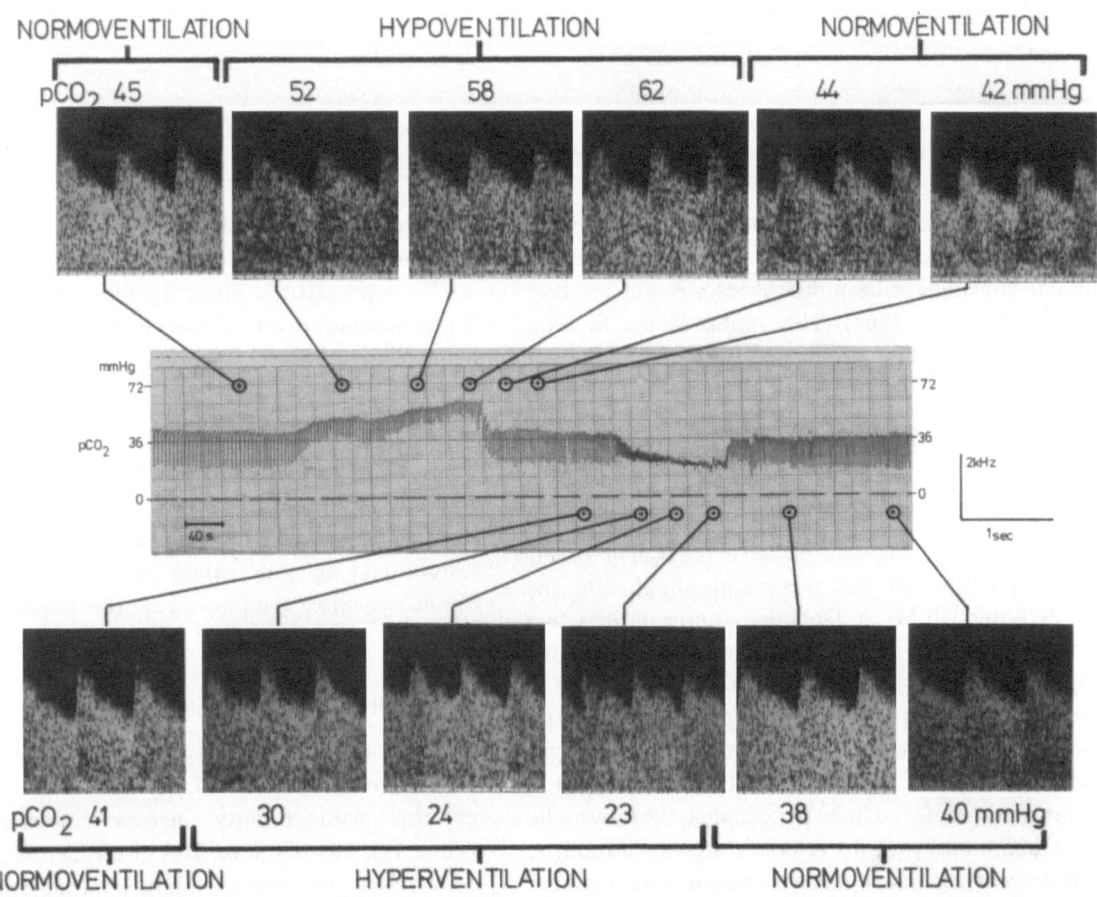

Fig. 104. CO$_2$ reactivity in a purely AVM supplying vessel (left PCA): almost no velocity changes occur with varying pCO$_2$

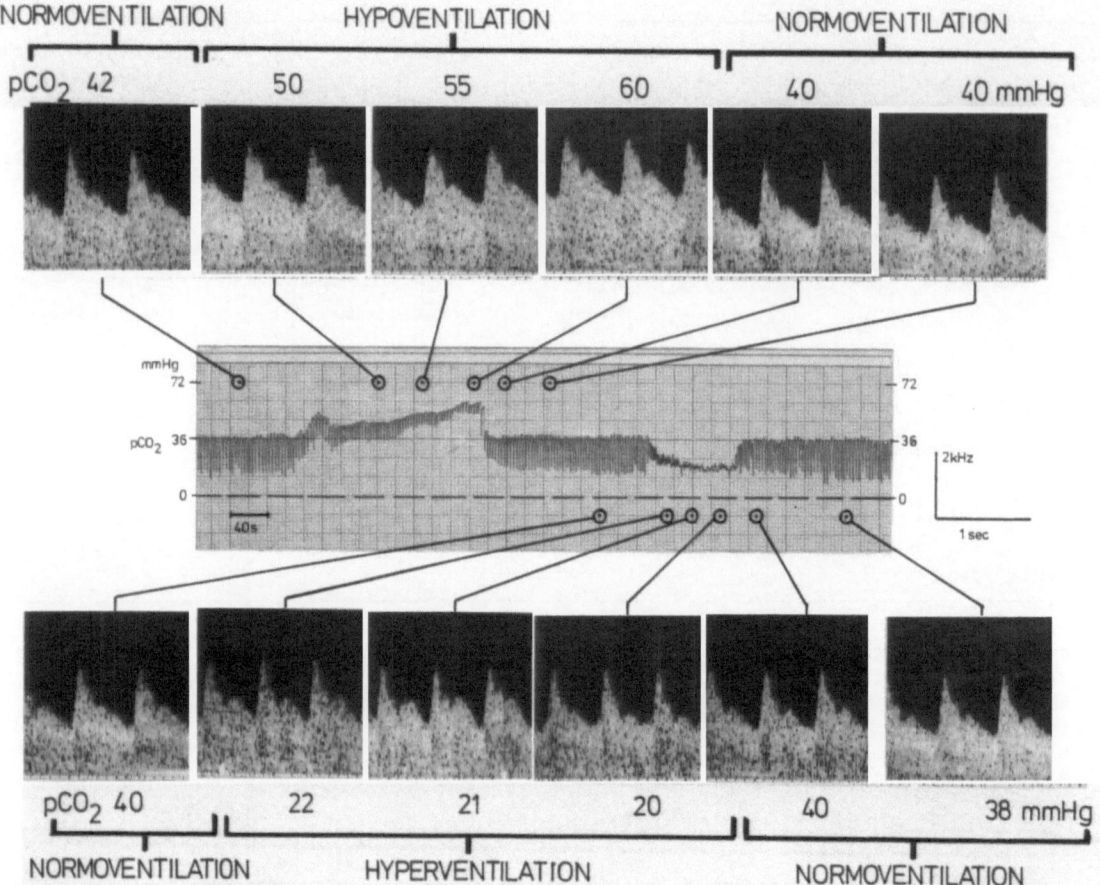

Fig. 105. CO_2 reactivity in a partially AVM supplying vessel (left MCA, same case as Fig. 104): slight flow acceleration in hypercapnia, slight systolic deceleration in hypocapnia but no diastolic hypocapnic response

Pure angioma feeders may show "steal phenomena" from normal brain in the case of large malformations. Similar effects have been described in recent cerebral infarcts (Brawley 1967). This report is the first on similar phenomena in angiomas. Further explanation and an impressive example are given in the following (Fig. 103).

CO_2 Reactivity and interhemispheric steal in angiomas

Fig. 103 shows an angioma located in the roof of the left ventricular trigonum that was mainly supplied by the left PCA and left MCA. Doppler patterns of these two vessels obtained in normocapnia are typical for angioma feeders, whereas contralateral MCA and PCA show normal spectra. With hypercapnia, flow velocities show normal increases in the brain supplying right MCA and right PCA. A very slight decrease is seen in the angioma feeders. Under hypocapnia, the flow in the right MCA and right PCA decelerates as normal. A paradoxical response is, however, found in the angioma feeding vessels which show increasing frequencies.

They are due to increasing stream resistance in normal brain supplying vessels and to the fact that angiomas are "vasoparalytic". Thus, the blood in the Circle of Willis becomes diverted into the malformation (interhemispheric steal, see Fig. 120, p. 107).

b) Mainly angioma supplying arteries

Cerebral arteries with most of their flow supplying the angioma show a systolic flow velocity of 3.5–4 kHz and a diastolic velocity of 2.5–3 kHz. In hypercapnia, systolic and diastolic values increase, whereas hypocapnic responses are poor. The total capacity of CO_2 reactivity is also poor and amounts to 1–2 kHz.

c) Partially angioma supplying arteries

Cerebral arteries with only partial contribution to an AVM show almost normal, pulsatile normocapnic flow patterns with slightly increased flow velocities (Table 19). Identification as AVM feeders is possible in comparison to corresponding contralateral vessels. Interhemispheric normocapnic differences, especially

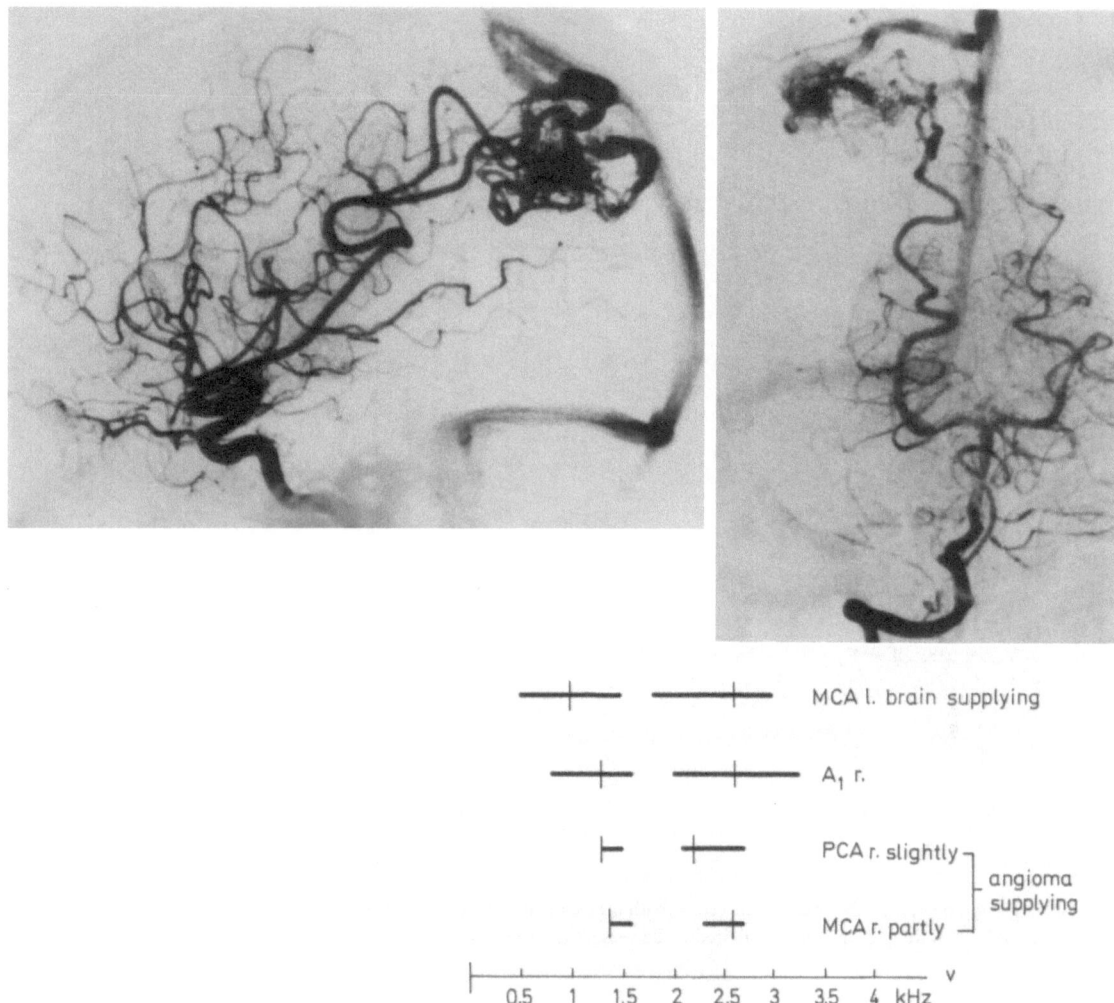

Fig. 106. Angiography and CO₂ reactivity of a partially AVM feeding (right MCA) and a slightly AVM feeding vessel (right PCA) in the case of a parietal angioma

in diastolic velocities, are above 0.5 kHz. Abnormal findings are, however, more distinct in tests of the CO₂ reactivity. Responses of the diastolic flow velocities are poor, especially in hypocapnia. Reactivity of the systolic velocity is normal, the total capacity is reduced to 2–3 kHz (Fig. 105).

The more distally an angioma feeder is evaluated, the more reduced is its regulative capacity (Fig. 107).

d) Cerebral arteries with only little contribution to an AVM

Such vessels supply mainly brain and to a lesser extent contribute to an angioma. Flow velocities and total capacity of CO₂ reactivity are within normal limits. In cases of basally located or large distal malformations, abnormalities can be found (Figs. 107, 108). Interhem-

ispheric difference may be above 0.5 kHz and the minimal hypocapnic diastolic velocity may stay above 1.0 kHz (normal: 0.5 kHz).

e) Cerebral arteries without contribution to an angioma and without impaired CO₂ reactivity

Total capacity of CO₂ reactivity in normal brain vessels of angioma patients is above 2.5 kHz. Diastolic and systolic responses are normal in most cases with hypocapnic diastolic deceleration down to 0.5 kHz. This even applied for most of our occipital angiomas, where both angioma-feeding and brain supplying PCA issue from the same vessel (basilar artery) and where, according to theoretical considerations, impaired autoregulation of the brain supplying branch had to be most readily expected.

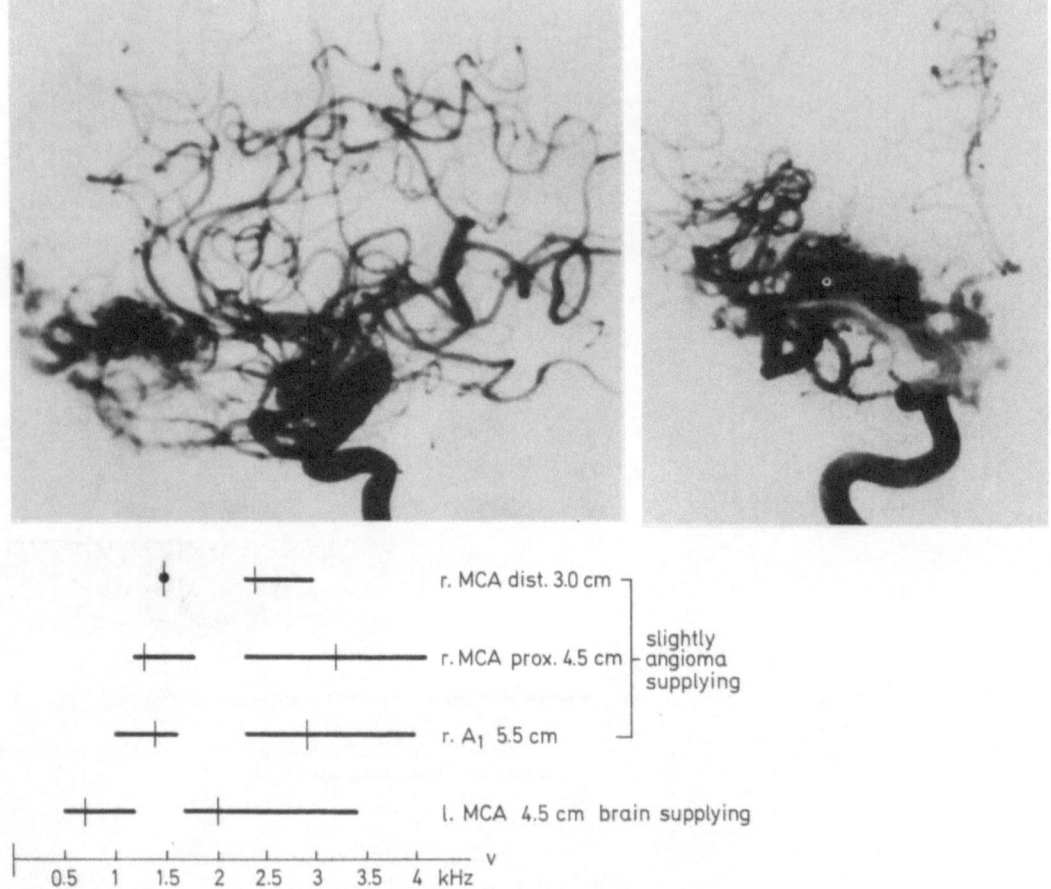

Fig. 107. Angiography and CO_2 reactivity of vessels with little contribution to a frontobasal AVM (right MCA, right ACA): the closer to the AVM the MCA is evaluated, the more impaired is the vessel's diastolic CO_2 response

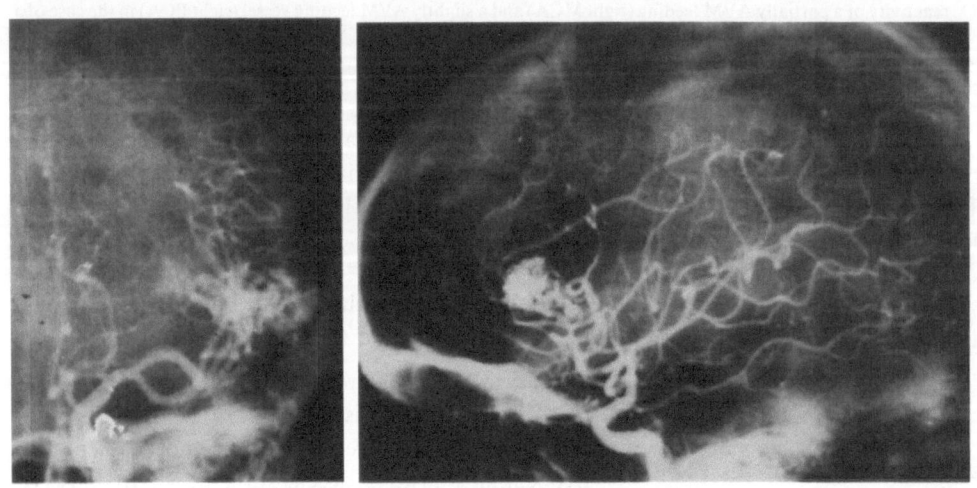

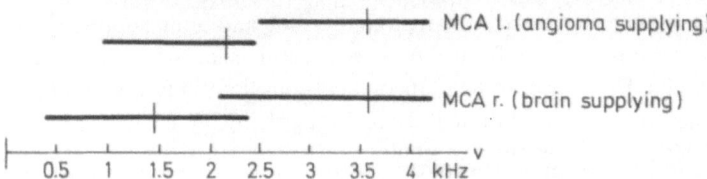

Fig. 108. Angiography and CO_2 reactivity in the case of a (proximal) left opercular angioma: The feeding left MCA shows only slight impairment of its diastolic CO_2 response

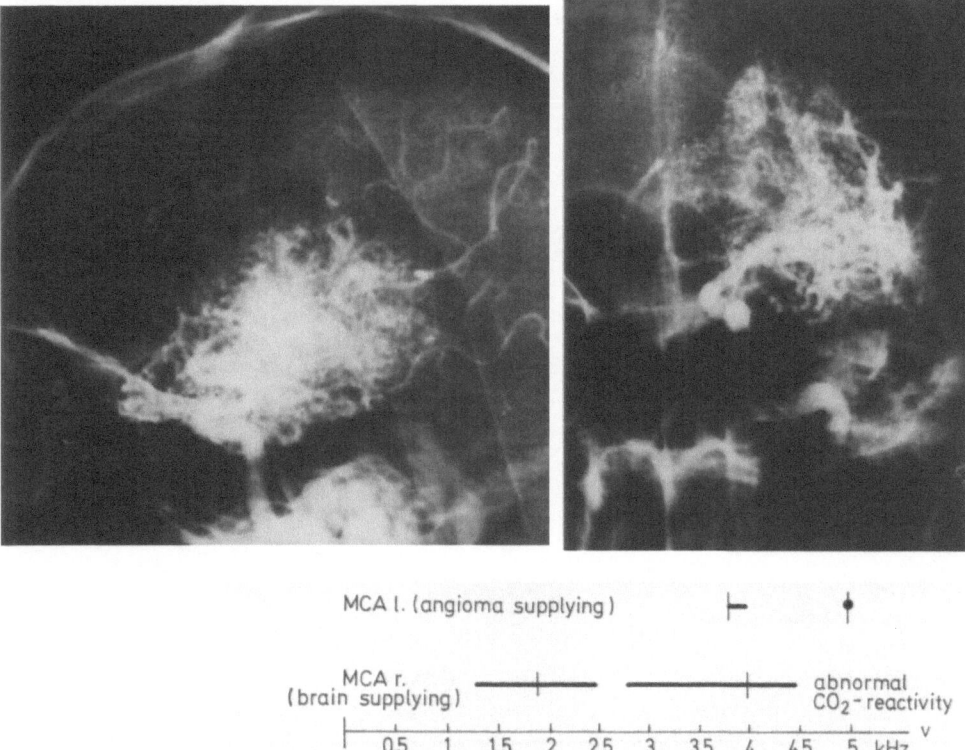

Fig. 109. Angiography and CO$_2$ reactivities in the case of a large Sylvian angioma: The AVM-feeding left MCA has very poor responses; the purely brain supplying right MCA shows an impaired diastolic flow deceleration under hypocapnia, minimal values reach only 1.3 kHz (normal: 0.5 kHz)

f) Cerebral arteries without contribution to an angioma but with impaired CO$_2$ reactivity

Only in large AVMs did purely brain supplying arteries show impaired CO$_2$ responses. Their patterns were similar to those of partial AVM feeders and characterized by impaired hypocapnic diastolic flow deceleration. An example is given in Fig. 109.

Fig. 109 shows a large left Sylvian angioma. In left carotid angiography, only the left MCA and AVM are visualized. In right carotid angiography, the right MCA, right ACA, left ACA, left MCA and angioma become visible (steal effect). Tests of the CO$_2$ reactivity in this case were abnormal even in the brain supplying right MCA. Hypocapnic diastolic velocities did not decrease below 1.3 kHz in this vessel (normal: 0.5 kHz) and thus indicated an impaired ability for arteriolar vasoconstriction in normal contralateral brain. This is most probably due to a reduced intravascular pressure in the right MCA.

3.5. CO$_2$ Reactivity of Brain Arteries in the Angiogram

In 2 angioma patients, angiography was performed under normocapnic, hypocapnic and hypercapnic conditions. We thereby tried to correlate blood flow velocity changes with morphological data. Angiographic perfusion time was measured (i.e., first visualization of carotid siphon until first visualization of parietal vein). We also observed changes of blood distribution under varying pCO$_2$.

Results (Table 20):

— Hypercapnia shortens the angiographic perfusion time of normal brain, whereas hypocapnia prolongs it considerably.
— Hypercapnia delays the angiographic filling of the angioma; hypocapnia does the opposite.
— Hypercapnia delays the angiographic drainage of the angioma. Hypocapnia has no effect in this regard.
— Contrast filling of the AVM is more intense in hypocapnia than in normocapnia (Figs. 111, 112).
— The diameter of the ICA siphon does not change with varying pCO$_2$.
— The proximal and distal diameters of AVM feeding arteries do not change with varying pCO$_2$.
— The diameters of small brain supplying arteries leaving the angioma feeders do change with varying pCO$_2$: They increase with hypercapnia and decrease with hypocapnia.
— Altered blood distribution was demonstrable in the

Table 20. CO_2 *Reactivity of Brain Vessels During Angiography in Angioma Patients (Example: Large Angioma of Ventricular Trigonum Supplied*

	CO_2 normal	$CO_2\uparrow$	$CO_2\downarrow$
Angiographical circulation time	5.5 sec	4.25 sec	8 sec
Time until filling of the first draining vein	1.75 sec	2.25 sec	1.5 sec
Time of visualisation of angioma nidus	4.75 sec	5.25 sec	4.75 sec
Diameter of ICA siphon	6.4 mm (no change with varying pCO_2)		
left MCA (main feeder) proximal	5.5 mm (no change with varying pCO_2)		
distal	3.0 mm (no change with varying pCO_2)		
distal	4.9 mm (no change with varying pCO_2)		
left MCA (opercular branches, brain supplying) proximal	1.8 mm	2.0 mm	1.7 mm
distal	1.0 mm	1.4 mm	0.8 mm

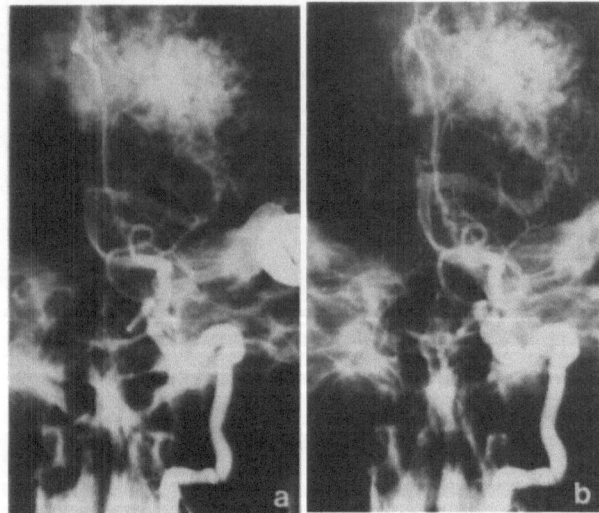

Fig. 110. Angiography in the case of an angioma of the left ventricular trigonum under varying pCO_2: a) Normocapnia ($pCO_2 = 40$ mmHg), 0.75 sec after dye injection, b) hypercapnia ($pCO_2 = 65$ mmHg), 0.75 sec after dye injection
The right PCA is only visualized under hypercapnia, due to arteriolar dilation in its peripheral territory. I am grateful to Prof. Schumacher and Dr. Bien, Neuroradiology of the University of Freiburg for the angiograms

case of an occipital angioma. Hypocapnia enhanced the visualization of the (vasoparalytic) malformation (Fig. 110), whereas the contralateral PCA was no longer visible. Hypercapnia reversed this effect; the normal PCA was now clearly visualized.

3.6. CO_2 Reactivity After AVM Removal

Consecutive tests of CO_2 reactivity after AVM removal were performed on 20 patients. In one patient, CO_2 reactivity was measured before and after partial embolisation as well as following surgical removal of the malformation.

a) Preoperatively exclusively or mainly AVM supplying arteries

These vessels had abolished or poor CO_2 reactivity prior to surgery. With at most 7 days following surgery, their reactivity normalized and became the same as in contralateral arteries. Hypocapnic diastolic flow de-

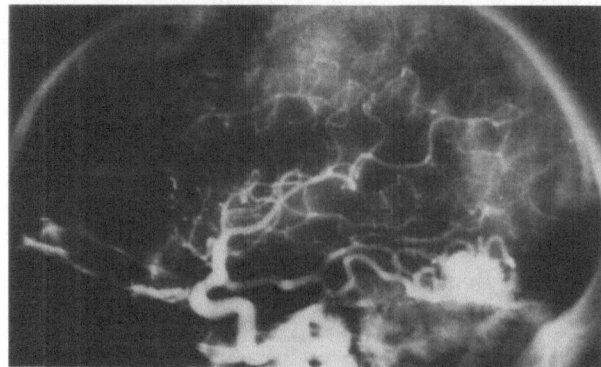

Fig. 111. Angiography in the case of a left temporooccipital AVM; normocapnia ($pCO_2 = 40$ mmHg); 1 sec after dye injection

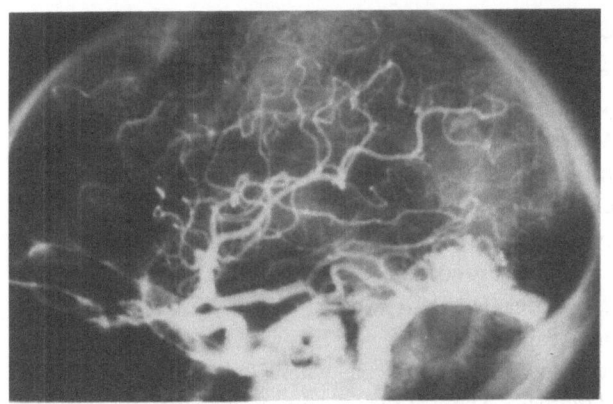

Fig. 112. Angiography in the same case as Fig. 111; hypocapnia ($pCO_2 = 20$ mmHg); 1 sec after dye injection: faster perfusion of the AVM in hypocapnia

celerations in the former feeders then reached 0.5 kHz; systolic and diastolic reactivities were within normal limits. One patient developed vasospasm on the seventh postoperative day. He had suffered from an intra-operative ventricular hemorrhage, but nevertheless maintained normal CO_2 reactivities, although on a higher level.

b) Preoperatively partially AVM supplying arteries (Fig. 113)

Vessels of preoperative partial AVM supply developed normal CO_2 reactivities within 4 days following surgery.

PREOPERATIVE STATUS

NORMOVENTILATION   HYPOVENTILATION   HYPERVENTILATION

POSTOPERATIVE STATUS

NORMOVENTILATION   HYPOVENTILATION   HYPERVENTILATION

MCA r.

PCA r.

A₁ r.

	PREOPERATIVE				POSTOPERATIVE		
	NORMO	HYPO	HYPER		NORMO	HYPO	HYPER
PH	7.395	7.312	7.579	PH	7.405	7.301	7.595
PCO2	41.8	54.3	24.8	PCO2	42.3	56.2	23.8
PO2	65.1	107.7	93.0	PO2	79.9	112.4	103.6
HCO3	25.2	26.8	23.3	HCO3	26.1	27.0	23.2
TCO2	26.4	28.4	24.0	TCO2	27.3	28.7	23.9
BE	.4	-000.1	3.1	BE	1.3	-000.0	3.1

2kHz

1sec

MCA l.

A₁ r.   slightly

PCA r.   mainly   angioma supplying

MCA r.   partly

MCA l.

A₁ r.

PCA r.

MCA r.

0.5 1 1.5 2 2.5 3 3.5 4 kHz

0.5 1 1.5 2 2.5 3 3.5 4 kHz

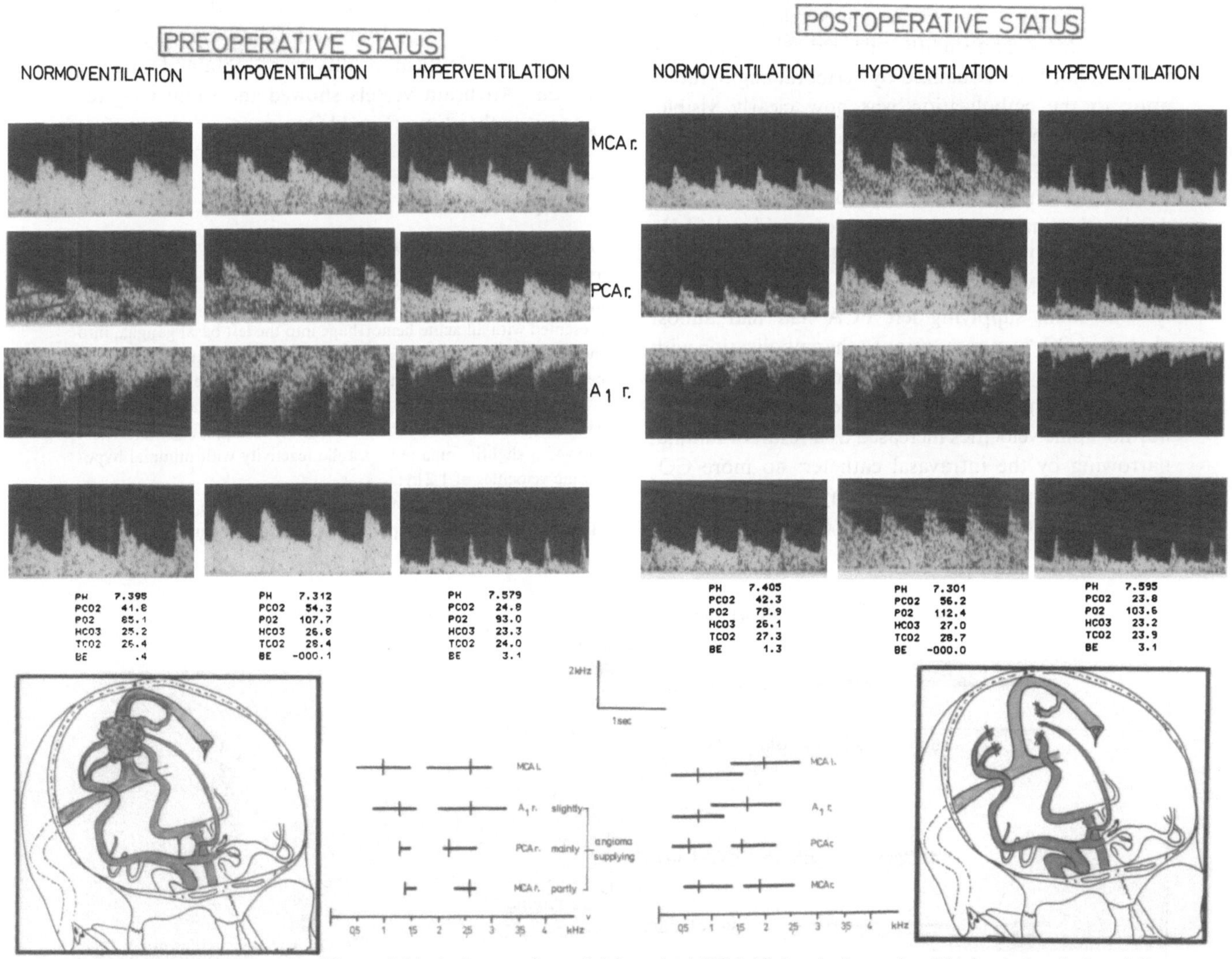

Fig. 113. Pre- and postoperative CO_2 reactivities in the case of a small right parietal AVM. All three brain arteries of this hemisphere had partially contributed to the angioma and developed normal CO_2 response within 4 days following surgery

c) Arteries with little preoperative contribution to an angioma

If the CO_2 reactivities in these vessels had been abnormal prior to the operation, they normalized within one day after AVM removal.

3.7. CO_2 Reactivity Before and After Partial Embolization

One patient had his left occipital angioma partially embolized by Professor Merland, Paris. A catheter was directed transcutaneously via the left ICA, left posterior communicating and the main feeding left posterior cerebral artery into the angioma. Large parts of the malformation were embolized with synthetic material. The catheter had to be left in place. Upon control angiography, the angioma had decreased in size. The supplying left MCA and left PCA still filled the malformation. The PCA diameter had decreased while left MCA had slightly enlarged. The right PCA that had not been visualized by vertebral angiography prior to the embolization was now clearly visible, indicating improved blood distribution.

The CO_2 reactivity of the left MCA had increased 4 weeks after embolization. Hypocapnic diastolic flow deceleration improved with velocities reaching 1.3 kHz compared to 1.7 kHz previously (Fig. 114). The normocapnic left MCA velocities had changed only slightly. The main supplying left PCA had had almost abolished CO_2 responses prior to the embolization with very slight hypercapnic flow accelerations and no hypocapnic change. Following the embolization, the normocapnic velocities increased as a result of luminal narrowing by the intravasal catheter: no more CO_2 response in hypercapnia, whereas hypocapnia now led to considerable flow decelerations (Fig. 114).

CO_2 Reactivity following surgical removal of the residual angioma

The same patient underwent surgical removal of his AVM 6 weeks after the embolization. The postoperative course was uneventful. The CO_2 reactivities were determined on the second and fifth postoperative day, after 4 weeks and 3 months (Fig. 115 a–f).

On the second day, CO_2 reactivities of the former main feeder (left PCA) were still poor (Fig. 115 e). The normocapnic diastolic velocity was low and did not decrease with hypocapnia, whereas in hypercapnia it did increase. The systolic velocity showed the same type of response. The left MCA, however, which had been a partial feeder showed almost normal CO_2 reactivities on the second postoperative day. Hypocapnic diastolic flow deceleration reached 0.8 kHz (normal: 0.5 kHz). At the fifth day, CO_2 responses of both vessels had returned to already normal values. Further improvement was found upon four-week follow-up (Fig. 115 e) and almost identical results were obtained after 3 months. All brain vessels showed the same CO_2 responses at that time (Fig. 116).

3.8. Special Cases

a) CO_2 Reactivity before and after hemorrhage from an AVM

This patient had a large left Sylvian angioma that was considered inoperable. One week after examination of CO_2 reactivities he presented with an acute hemorrhage into the left basal ganglia, into the ventricle and subarachnoid space. Follow-up examinations were performed 21 days, 1 month and 5 months after the hemorrhage.

Prior to the bleeding, CO_2 responses of the main feeding left MCA had been almost abolished. The brain-supplying right MCA showed a slightly impaired diastolic reactivity with minimal hypocapnic velocities of 1.2 kHz.

At the twenty-first day following the hemorrhage, we found signs of vasospasm in vessels supplying brain and angioma. Systolic and

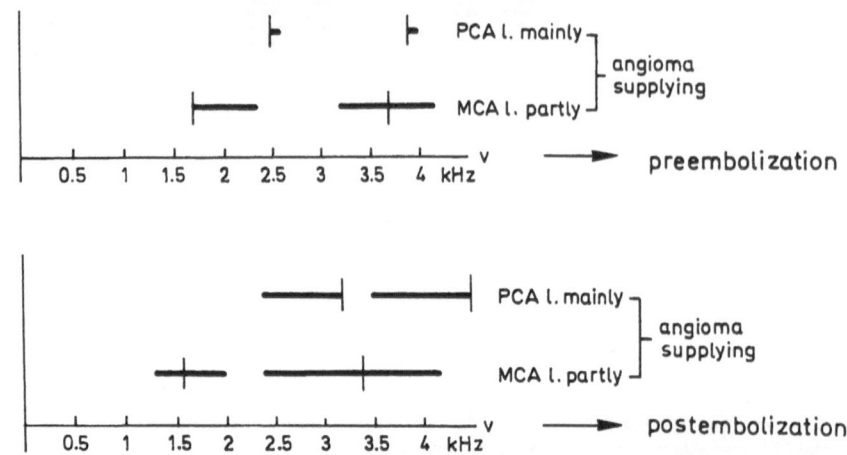

Fig. 114. CO_2 reactivities in left PCA and left MCA before and after partial embolization (see text)

regulatory capacity
of normal and angioma supplying vessels

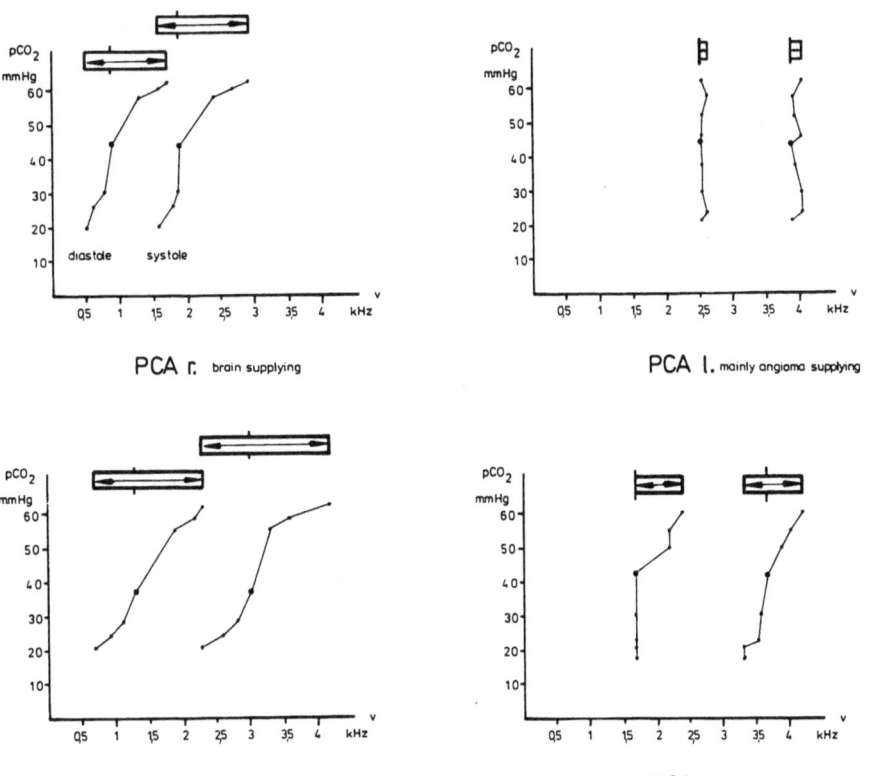

regulatory capacity
of normal and angioma supplying vessels

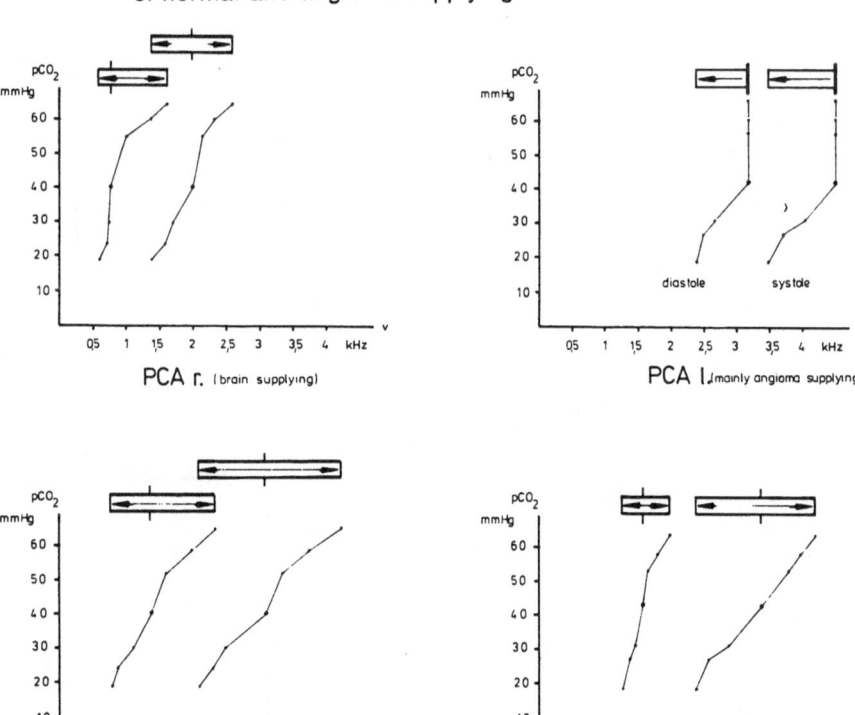

Fig. 115. CO_2 reactivities in the case of a left occipital angioma supplied by left PCA and left MCA: a) prior to embolization, b) after partial embolization, c) two days after surgical removal, d) five days after surgery, e) four weeks after surgery, f) three months after surgery

regulatory capacity
of brain vessels after removal of an occipital angioma

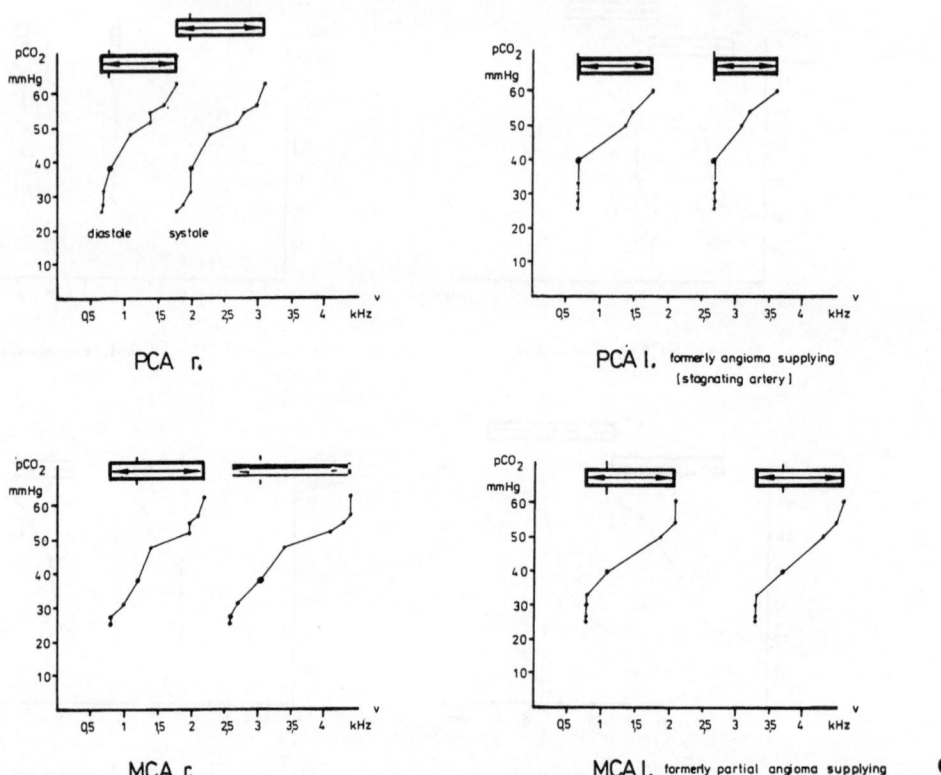

PCA r.

PCA l. formerly angioma supplying
(stagnating artery)

MCA r.

MCA l. formerly partial angioma supplying   **c**

regulatory capacity
of brain vessels after removal of an occipital angioma

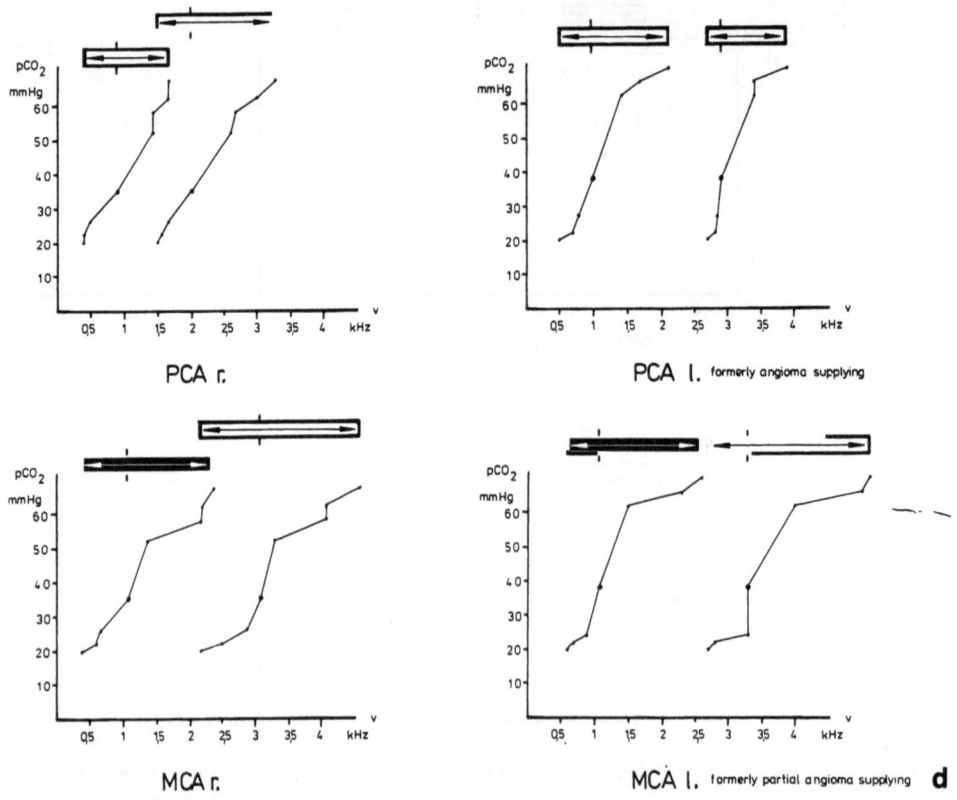

PCA r.

PCA l. formerly angioma supplying

MCA r.

MCA l. formerly partial angioma supplying   **d**

regulatory capacity
of brain vessels after removal of an occipital angioma

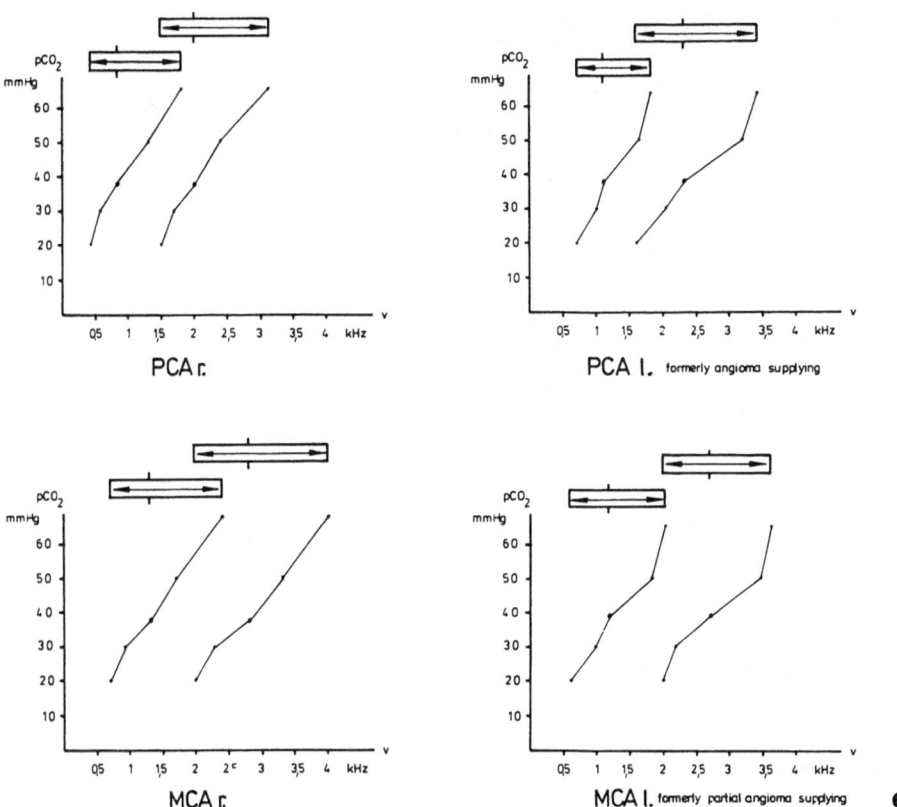

PCA r.

PCA l. formerly angioma supplying

MCA r.

MCA l. formerly partial angioma supplying   e

regulatory capacity
after removal of l. occipital angioma

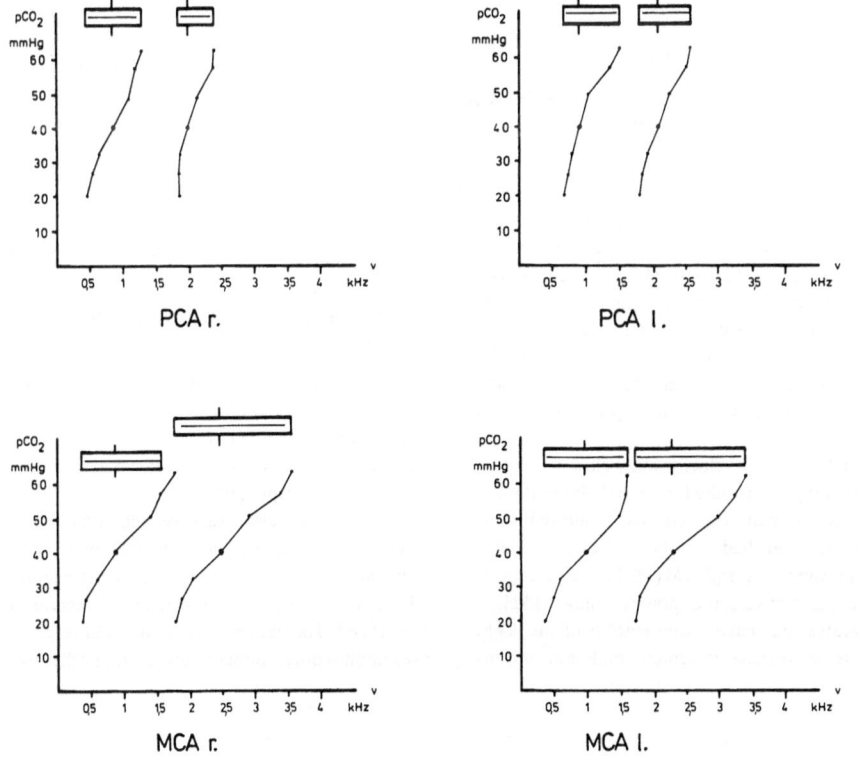

PCA r.

PCA l.

MCA r.

MCA l.   f

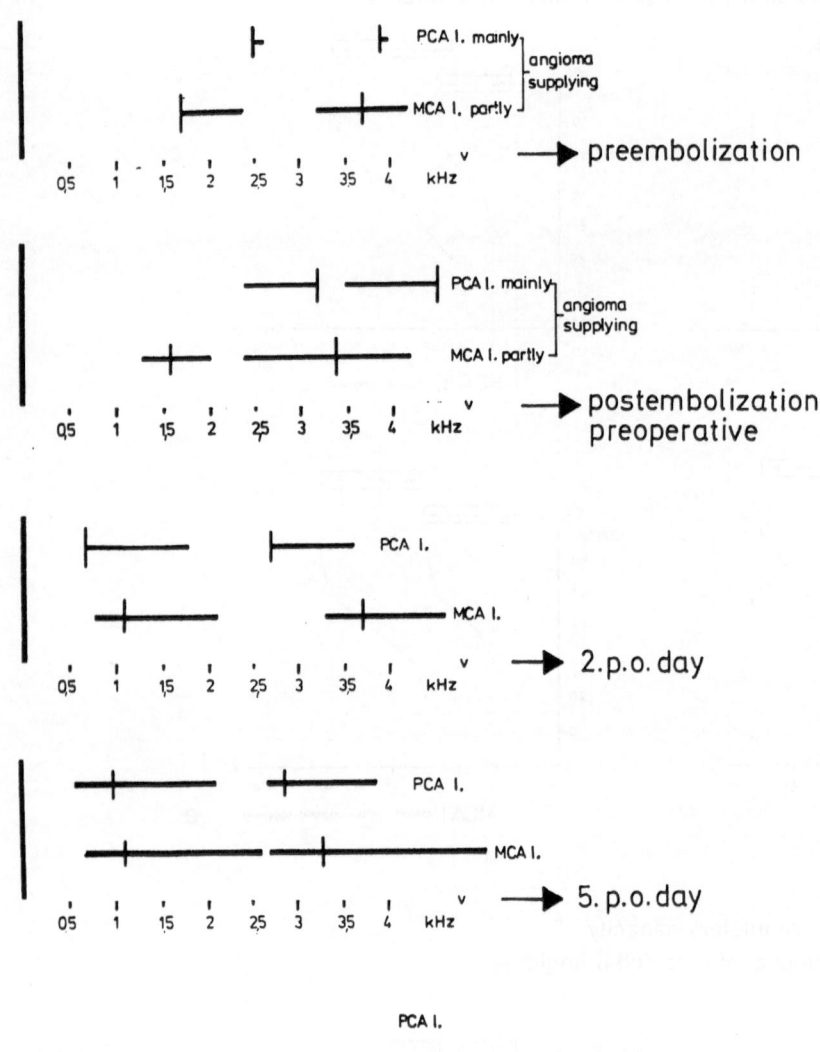

Fig. 116. Synopsis of the CO_2 reactivity changes in feeders of the embolized and later removed occipital AVM (see text)

diastolic velocities were higher. The diastolic CO_2 reactivity of the left MCA was still abolished. The systolic flow, however, decelerated under hypocapnia and hypercapnia. The right MCA showed reduced CO_2 reactivities when compared to our initial findings. After 4 weeks, with subsiding vasospasm, flow velocities had decreased (Fig. 113). CO_2 reactivity of the left MCA now was the same as before the hemorrhage. The right MCA had regained its reactivity. Normocapnic velocities in this vessel were still slightly faster than initially.

After 5 months, left MCA flow velocity was below the intitial value. Obviously, the bleeding had resulted in partial obliteration of the AVM. Corresponding to that, the left MCA showed some hypercapnic CO_2 response that had not been present initially. Responses of the brain supplying right MCA had also slightly improved with minimal diastolic velocities now reaching 1.1 kHz.

This case demonstrates that partial obliteration of an AVM following a spontaneous hemorrhage is demonstrable not only by improvement of the normocapnic flow velocities, but also by improvement of CO_2 responses in vessels supplying brain and AVM.

b) CO_2 Reactivity in clinical brain death syndrome following AVM removal (Figs. 118, 119)

This patient had an infratentorial angioma that was supplied by both superior cerebellar and posterior inferior cerebellar arteries. In the 2 days following complete removal, three hemorrhages occurred and were each evacuated. A ventricular drainage was applied because of occlusive hydrocephalus.

On the day after surgery and after evacuation of the first cerebellar hemorrhage, transcranial Doppler evaluation showed slow flow, early diastolic zero flow and high peripheral stream resistance in all basal arteries. Intraventricular pressure at that time was 50 cm H_2O. The patient was on artificial respiration but opened his eyes upon request. Some hours later, another two hemorrhages had

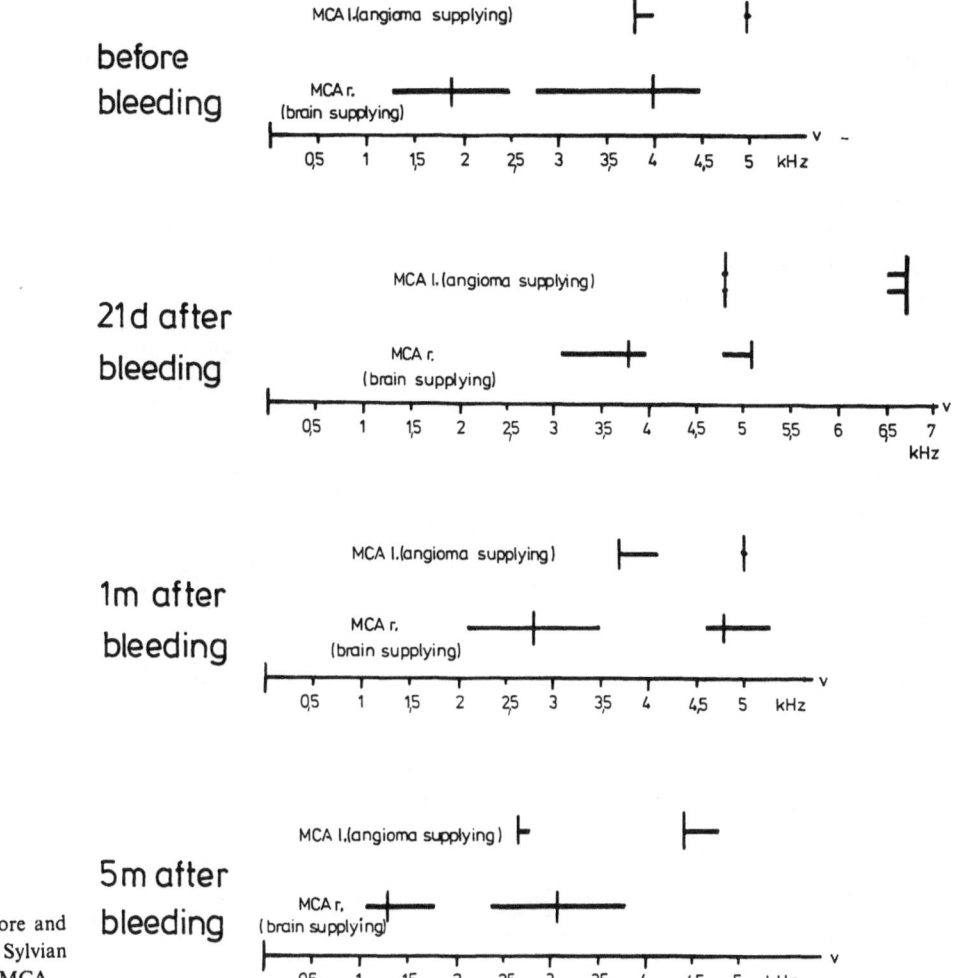

Fig. 117. CO_2 reactivities before and after bleeding from a left Sylvian angioma supplied by the left MCA

to be evacuated. On the second postoperative day, the patient was comatous without spontaneous respiration. During tanscranial Doppler examination, the mean systemic arterial blood pressure was 41 mmHg. Systolic flow velocities were high, diastolic velocities were low due to high peripheral resistance. On the third postoperative day, with the clinical situation unchanged, the flow velocities had increased. In the MCA they had almost doubled; mean arterial blood pressure was 70 mmHg at that time. These pressure-dependent velocity changes indicate loss of cerebrovascular autoregulation.

On the fifth postoperative day, all clinical signs of brain death were present; arterial blood pressure could no longer be measured. In MCA and left PICA, only small systolic peaks were visible. The vertebral arteries showed oscillating flow, indicating cerebral circulatory arrest.

At the time of brain death, transcranial Doppler evaluation and CO_2 reactivity testing was done under SABP of 97/56 mmHg. The left MCA showed patterns typical of increased intracranial pressure. Systolic peaks were steep and rounded off; diastolic flow was very low. In hypercapnia, diastolic and systolic velocities decreased, which is a paradoxic response indicating complete vascular dysregulation. Hypocapnic flows also decelerated with diastolic values reaching almost zero. The left PICA showed a similar pattern. Flow velocities decreased in hypocapnia and in hypercapnia.

3.9. Discussion

Studies on cerebrovascular autoregulation have been conducted with very different methods (Table 21). Florey (1925) and Fog (1934) were the first to demonstrate autoregulation in response to blood pressure changes. Through "cranial windows" applied in cats, they observed dilation and constriction of the pial vessels with varying blood pressure. Lassen (1959) and Sokoloff (1960) later studied the limits of cerebral vascular autoregulation. Blood flow rates remained constant throughout a range from 50–130 mmHg mean systemic arterial pressure.

Besides blood pressure, release of CO_2 from the brain tissue is of importance in the regulation of arteriolar diameters. Cow (1911, *in vitro* study) even found the carotid artery to dilate in response to CO_2. Forbes (1928) and Wolff (1930) made the same observations in pial vessels of animals (cranial windows). Gibbs (1933), Schmidt and Hendrix (1937, thermo-

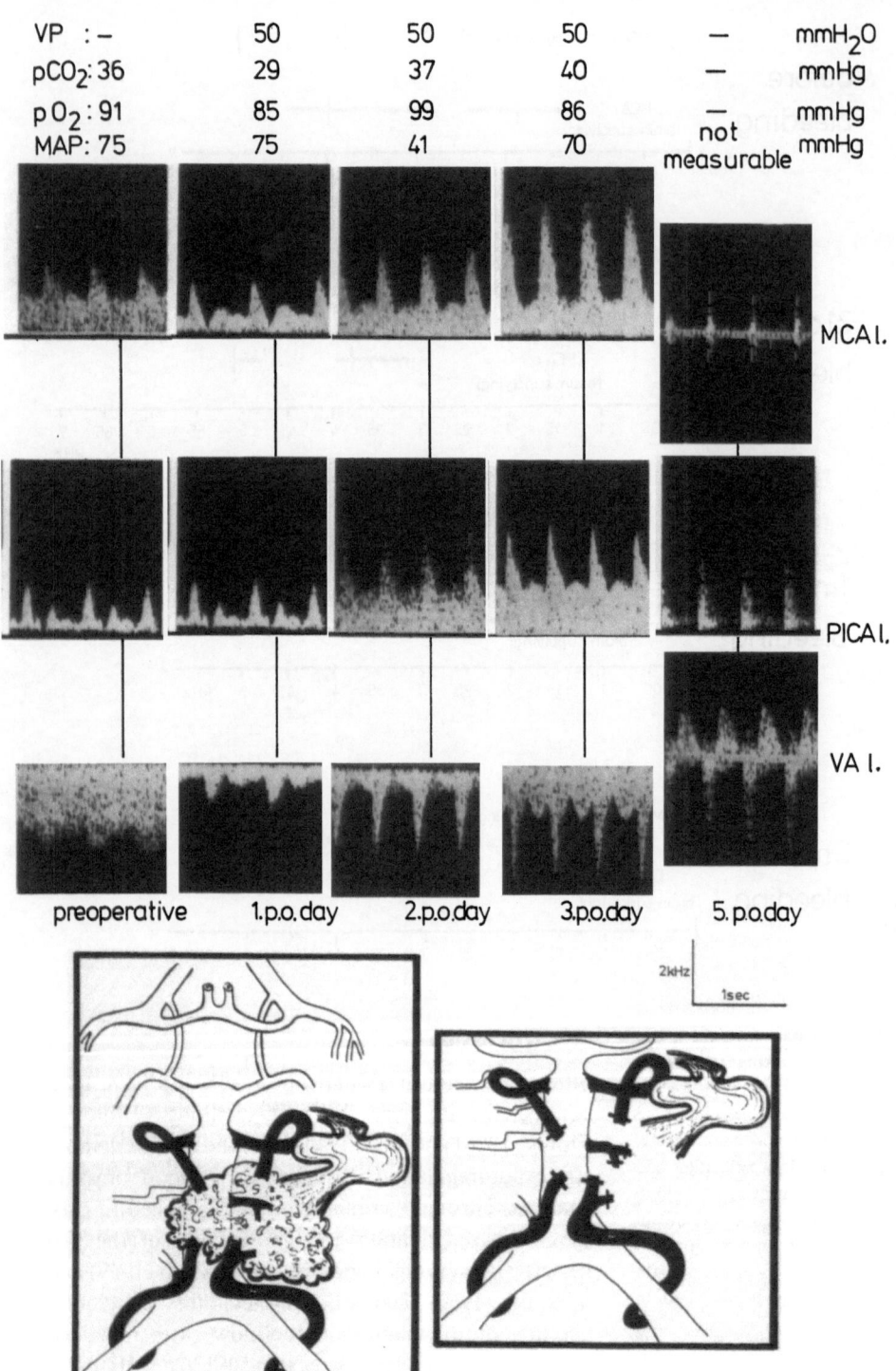

VP	: –	50	50	50	–	mmH$_2$O
pCO$_2$	: 36	29	37	40	–	mmHg
pO$_2$	: 91	85	99	86	–	mmHg
MAP	: 75	75	41	70	not measurable	mmHg

MCA l.

PICA l.

VA l.

preoperative   1.p.o.day   2.p.o.day   3.p.o.day   5.p.o.day

2kHz

1sec

Fig. 118. Changes of flow velocities in the case of an infratentorial angioma with postoperative central dysregulation and brain death

probes) showed such vascular changes to correspond with changes of the tissue's blood flow. Kety and Schmidt (1946, N$_2$O method) were the first to quantify such phenomena, although measuring times with the NO$_2$ method are too long (10–14 min.) to provide a precise monitoring. Radioactive wash-out studies were more useful in this regard and also allowed regional measurements (Lewis 1960, Lassen and Ingvar 1961, Obrist 1975, Tominaga 1976 and Lassen 1972). The maximal range of CO$_2$-dependent cerebrovascular regulation was studied by Reivich (1964, animals). He experimentally varied the arterial pCO$_2$ from 5 to

Table 21. *Methods for Measurement of Cerebral Perfusion and Autoregulation*

Site of measurement	Method	Authors
Brain tissue	thermoprobe	Schmidt & Hendrix 1937
		Reivich 1964
		Brawley 1968
		Betz 1965
		Wüllenweber 1965
	nitrogen oxide method	Kety & Schmidt 1946
		Shalit 1967
		Smith 1971
	O_2-difference	Shapiro 1965
	radioalbumin	Fieschi 1963
	krypton[85]	Harper 1965
		Waltz 1970
		Ekstrom 1971
		Lassen 1961
	133 xenon	Lassen & Ingvar 1961
		Palvölgyi 1969
		Tsuda 1983
	regional 133 xenon	Lassen & Ingvar 1972
		Yamamoto 1971
	hydrogen method	Meyer 1972
Brain vessels cortical basal	fluorescein	Feindel 1971
	angiographically	Greitz 1956
	Doppler	Strassburg 1982
		Batton 1982
		Bada 1982
		Jorch 1985
	transcranial Doppler	Aaslid 1982
Cervical carotid	electromagnetically	Sagawa & Guyton 1961
	Doppler	Thoresen 1979
		Beasley 1979
		Hanga 1980

418 mmHg and found blood flow changes to occur throughout the range from 15 to 150 mmHg pCO_2. Values below or above these limits did not cause further flow alterations. CO_2 responses are impaired under conditions such as hypertonus, hypotonus (Harper 1965, Ekstrom 1971, Raichle 1972, Artu 1985) and severe hypoxia (Reivich 1964). Other factors such as neuronal influence (Sokoloff 1960) and pO_2 changes between 50 and 300 mmHg (Reivich 1964, Lambertsen 1961) are of less importance.

In summary, the following concepts on cerebral blood flow regulation are generally accepted today:

—Autoregulation in response to blood pressure is effective throughout a range of 50–150 mmHg systemic arterial mean blood pressure.

—Proportional relationship between cerebral blood flow and arterial pCO_2.

—Correlation between regional cerebral blood flow and regional metabolic activity.

—Lactate and adenosine serve as mediator substances.

Arteriolar diameter changes are a result of physical as well as chemical mechanisms:

—Physical factors: myogenic response of muscle cells to changes of the transmural pressure gradient (Bayliss effect, Lassen 1964, Ekstrom 1971).

—Metabolic factors: extracellularly dissolved CO_2 and

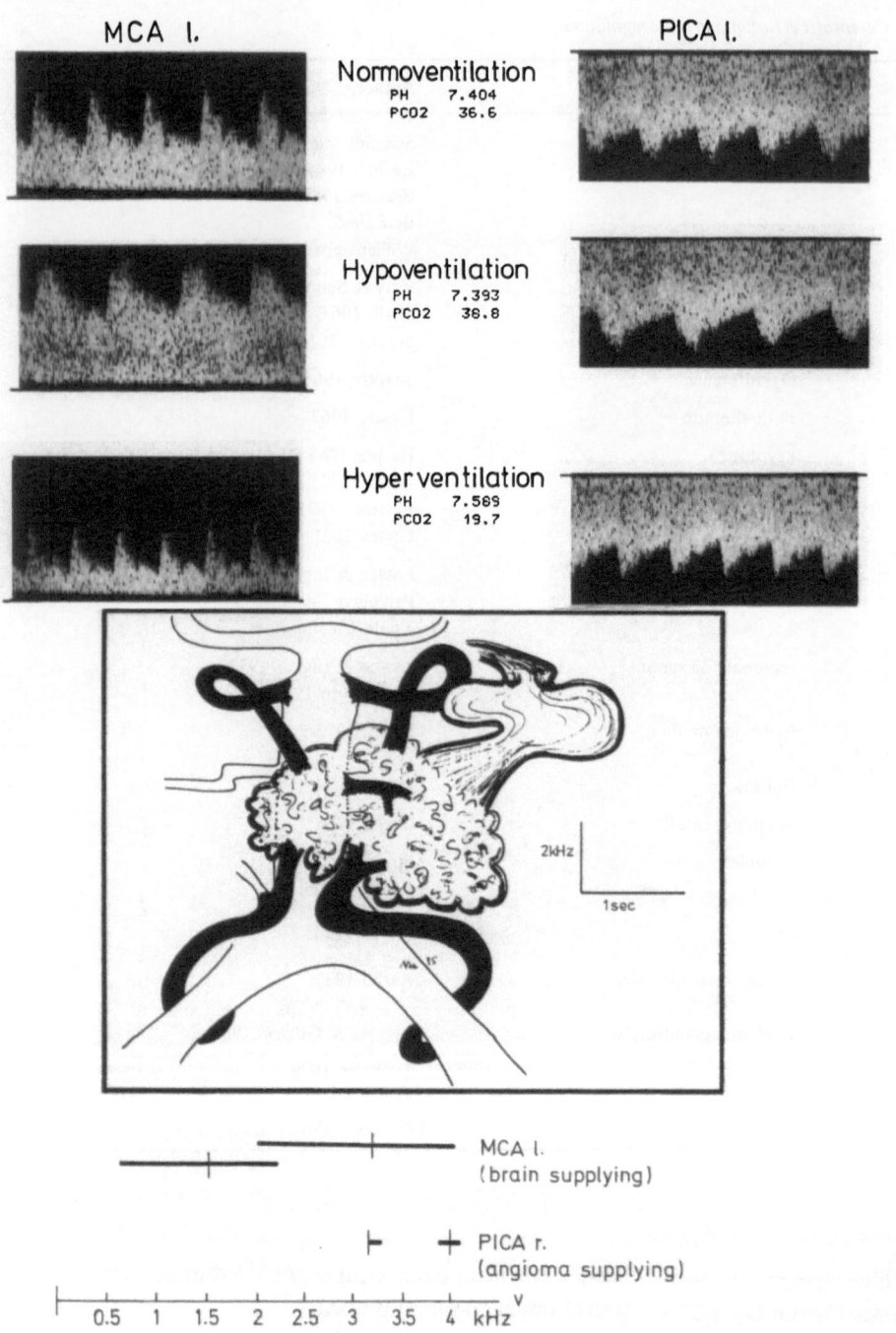

Fig. 119. CO_2 reactivities in a lethal case of an infratentorial angioma supplied by both PICA and both SCA: a) preoperative recording, b) postoperative recordings in clinical brain death showing paradoxical CO_2 responses (see text)

CSF bicarbonate influence the intracellular pH of the muscle cell which is relevant for constriction or relaxation (Gotoh 1961, Lassen 1966).

a) Methods

Cerebral blood flow can be measured directly and indirectly in the brain tissue itself or in the brain supplying vessels (Table 21). Tests of the CO_2 reactivity require methods that immediately detect occurring flow alterations. Transcranial Doppler sonography is very suitable in this regard. As the basal cerebral arteries do not react to pCO_2 changes (Huber 1967), the Doppler findings directly reflect diameter changes of the arteriolar vascular bed. Methodological errors due to unknown recording angles are slight.

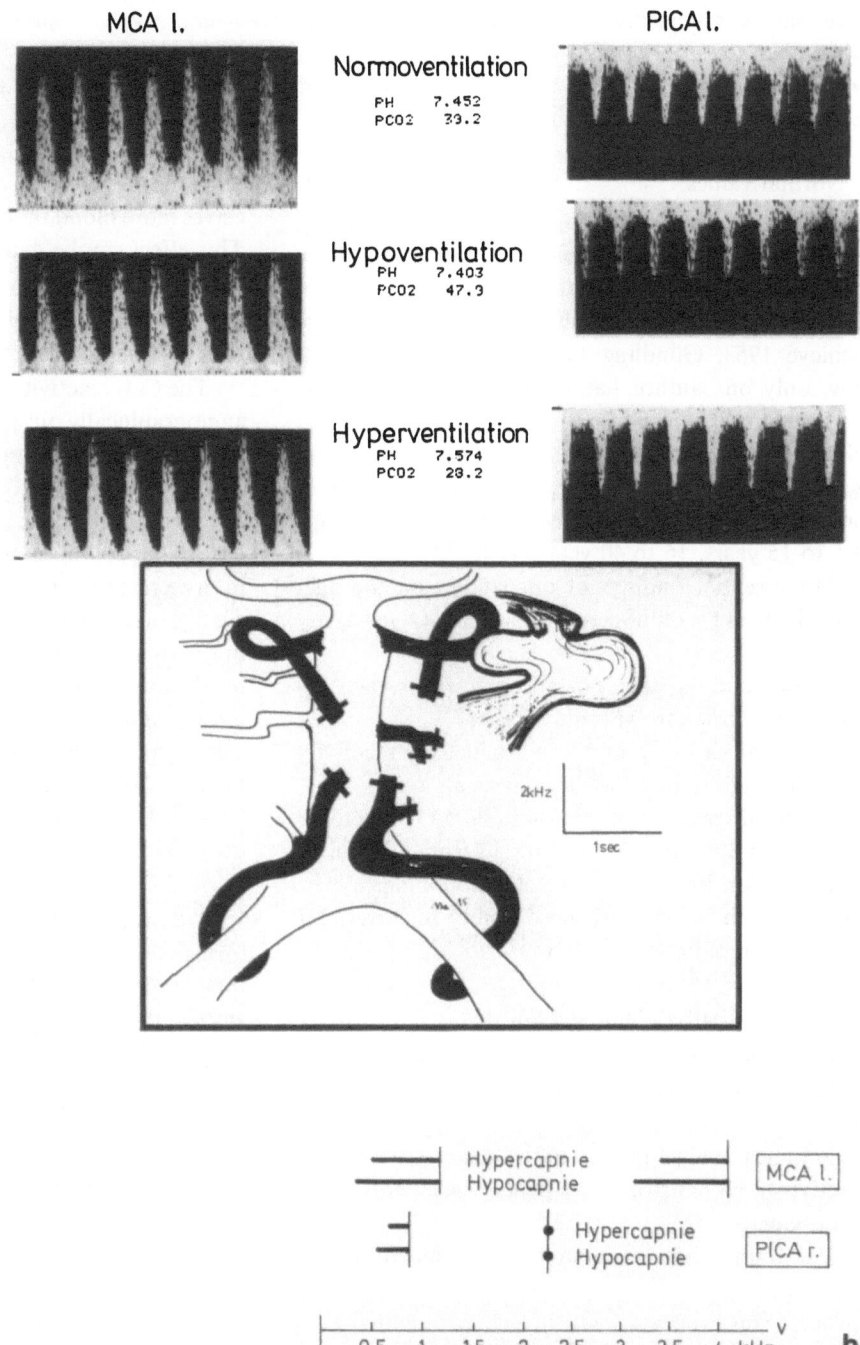

All former authors have induced hypocapnia by spontaneous hyperventilation. Different approaches have been described to induce hypercapnia. Simply holding one's breath is unsuitable since the intrathoracic and intracranial venous pressure rises and consequently may alter cerebral perfusion (Beasley 1979). We therefore chose the direct application of CO$_2$ to the inhaled air and waited for the endexpiratory pCO$_2$ to adjust to the desired hypercapnic values. This technique is more appropriate than the application of fixed air-CO$_2$ mixtures, as they may lead to considerable interindividual variations of pCO$_2$, depending on body weight and age (Markwalder 1984). Our measurements were documented each time a certain pCO$_2$ steady state had been maintained over at least 2 minutes. We monitored the endexpiratory pCO$_2$ values by infrared analysis;

these values reportedly provide a reliable indirect indicator of the arterial pCO_2 (Whiteshell 1981, Burki 1983). Transcutaneous pCO_2 recordings were not used because of their poor reliability.

b) Normal values

Former authors have described their CO_2 studies using radioalbumine, N_2O or xenon[133] techniques. They found the cerebrovascular CO_2 response to be preserved but slightly diminished with increasing age (Schieve 1953, Gündling 1985, Fieschi 1963). Until now, only one author has reported on the dopplersonographic CO_2 response of basal cerebral vessels in different age groups (Markwalder 1984). He found no age-dependent changes in 31 patients, perhaps due to the wide range of ages in each of his groups (age groups of 5 to 15 years, 16 to 40 years, 41 to 73 years).

The essential findings of our studies on the age-dependent course of the cerebrovascular CO_2 reactivity are:
—Systolic and diastolic normocapnic flows each decelerate by one-third with during life.
—The total capacity of the systolic CO_2 reactivity decreases slightly by 8% with increasing age.
—The total capacity of the diastolic CO_2 reactivity decreases by 32%; this decrease results from a diminishing ability for diastolic flow acceleration under hypercapnia, which indicates reduced ability for arteriolar relaxation in older people.

c) CO_2 Reactivity in angioma patients

CBF in angioma patients has been studied by several authors, although without significant results (Yamada 1982, Lassen 1964, Okabe 1983, Häggendal 1965, Oeconomos 1969, Prosenz 1971, Menon 1979). CO_2 reactivity in angioma patients has not been investigated.

We found that angioma feeding vessels can be identified not only by characteristic normocapnic flow patterns, but even more sensitively by impaired CO_2 responses. Hemodynamic involvement of "normal" brain arteries is sometimes demonstrable as well. Using these criteria, we propound a classification according to hemodynamic characteristics. Vessels can be described dopplersonographically as
—purely AVM supplying;
—mainly AVM supplying;
—partially AVM supplying;
—little AVM supplying;
—non-AVM supplying without hemodynamic involvement;
—non-AVM supplying with hemodynamic involvement.

Intracerebral steal phenomena in angioma patients are discernable in cases of large AVM (Fig. 120). While normal brain arterioles constrict in hypocapnia, the angioma lacks such vasomotricity and therefore receives more blood than under normocapnic conditions. This effect can be demonstrated angiographically and also by Doppler sonography, where the feeding vessels, in contrast to normal brain arteries, show increasing flow velocities under hypocapnia.

The CO_2 reactivity of brain vessels has been studied angiographically only in healthy subjects (Huber 1976) and brain tumor patients (Krueger 1963, Amundsen 1966, Rockloff 1966, Grange 1969, Du Bonlay 1970). These authors reported the angiographic perfusion time to be prolonged in hypocapnia and to be shortened in hypercapnia. Only small vessels were found to change in diameter. Steal effects have been described in brain tumors (Palvölgyi 1969, CBF measurements), head trauma (Wüllenweber, 1965, heat clearance technique) and cerebral infarcts (Brawley 1967). Brain tumors as well as angiomas feature a local impairment of vascular autoregulation, so that hypocapnia enhances their angiographic visualization. Hypercapnia does the opposite and leads to improved filling of normal vessels.

Following AVM removal, CO_2 responses of the former feeding arteries quickly return to normal. The time needed for this adjustment depends mainly on the vessel's previous degree of contribution to the malformation. Arteries with little AVM supply normalize immediately following the operation, whereas the angioma's main feeders may need up to 7 days. Such main feeding vessels have been described as postoperatively "stagnating" arteries because of their large diameters and extremely slow blood flow. The postoperative Doppler follow-up of CO_2 reactivities in stagnating vessels reveals characteristic findings: no response in hypocapnia for up to 7 days, normal flow acceleration in hypercapnia from the first day on. The hypocapnic response initially lacking has to be explained by the already maximal postoperative vasoconstriction in the peripheral distribution of the former main feeder. Further flow deceleration is impossible in this situation (see also Chapter V, Intraoperative measurements).

Brain arteries with or without previous hemodynamic involvement showed immediately normal CO_2 reactivities in any case following the removal of angiomas.

Table 22. CO_2 *Reactivity in Diseases Affecting the Brain*

Hypertension	$CO_2 \uparrow$ (man) $\rightarrow$ CBF $\uparrow$	(Fazekas 1953)
	$CO_2 \uparrow$ (ape) $\rightarrow$ CBF $\uparrow$	(Raichle 1972)
Hypotension	$CO_2 \uparrow$ (dog) $\rightarrow$ CBF $\leftrightarrow$	(Harper 1965)
	$CO_2 \uparrow$ (dog) $\rightarrow$ CBF $\leftrightarrow$	(Artu 1985)
Atherosclerosis	$CO_2 \uparrow$ (man) $\rightarrow$ CBF $\uparrow$	(Fazekas 1953)
Head trauma	$CO_2 \uparrow$ (man) $\rightarrow$ CBF $\leftrightarrow$	or $\downarrow$
	$CO_2 \downarrow$ (man) $\rightarrow$ CBF $\leftrightarrow$	or $\uparrow$ (Enevoldsen 1978)
	$CO_2 \downarrow$ (man) $\rightarrow$ CBF $\downarrow$	"acute" (Wüllenweber 1968)
Brain tumor (peritumoral tissue)	$CO_2 \uparrow$ (man) $\rightarrow$ CBF $\downarrow$	"intracerebral steal" (Palvölygi 1969)
	$CO_2 \downarrow$ (man) $\rightarrow$ CBF $\uparrow$	"inverse intracerebral steal"
	$CO_2 \downarrow$ (man) $\rightarrow$ CBF $\downarrow$	"intracerebral steal" (Wüllenweber 1968)
Cerebrovascular disease		
acute ischemia	$CO_2 \uparrow$ (man) $\rightarrow$ CBF $\rightarrow$	varying response (Fieschi 1963)
	$CO_2 \uparrow$ (dog) $\rightarrow$ CBF $\downarrow$	"intracerebral steal" (Brawley 1967) (I.C.S.)
	$CO_2 \uparrow$ (man) $\rightarrow$ CBF $\downarrow$	(I.C.S.) (Hoedt-Ramensen 1967)
	$CO_2 \uparrow$ (man) $\rightarrow$ CBF $\downarrow$	(I.C.S.) (Symon 1968, Lassen 1966)
	$CO_2 \uparrow$ (cat) $\rightarrow$ CBF $\leftrightarrow \downarrow$	(I.C.S.) (Waltz 1970)
	$CO_2 \uparrow$ (dog) $\rightarrow$ CBF $\uparrow$	"inverse intracerebral (Brawley 1967) steal" (I.I.C.S.)
	$CO_2 \downarrow$ (man) $\rightarrow$ CBF $\uparrow$	(I.I.C.S.) (Lassen 1966)
chronic ischemia	$CO_2 \uparrow$ (cat) $\rightarrow$ CBF $\uparrow$	(Waltz 1970) (after 12 days)
	$CO_2 \uparrow$ (man) $\rightarrow$ CBF $\uparrow$	(Paulsen 1970)
	$CO_2 \uparrow$ (man) $\rightarrow$ CBF $\uparrow$	(Meyer 1972)
	$CO_2 \downarrow$ (man) $\rightarrow$ CBF $\downarrow$	(Meyer 1972)
	$CO_2 \uparrow$ (man) $\rightarrow$ CBF $\uparrow$	(Tsuda 1983)
	$CO_2 \downarrow$ (man) $\rightarrow$ CBF $\downarrow$	(Tsuda 1983)

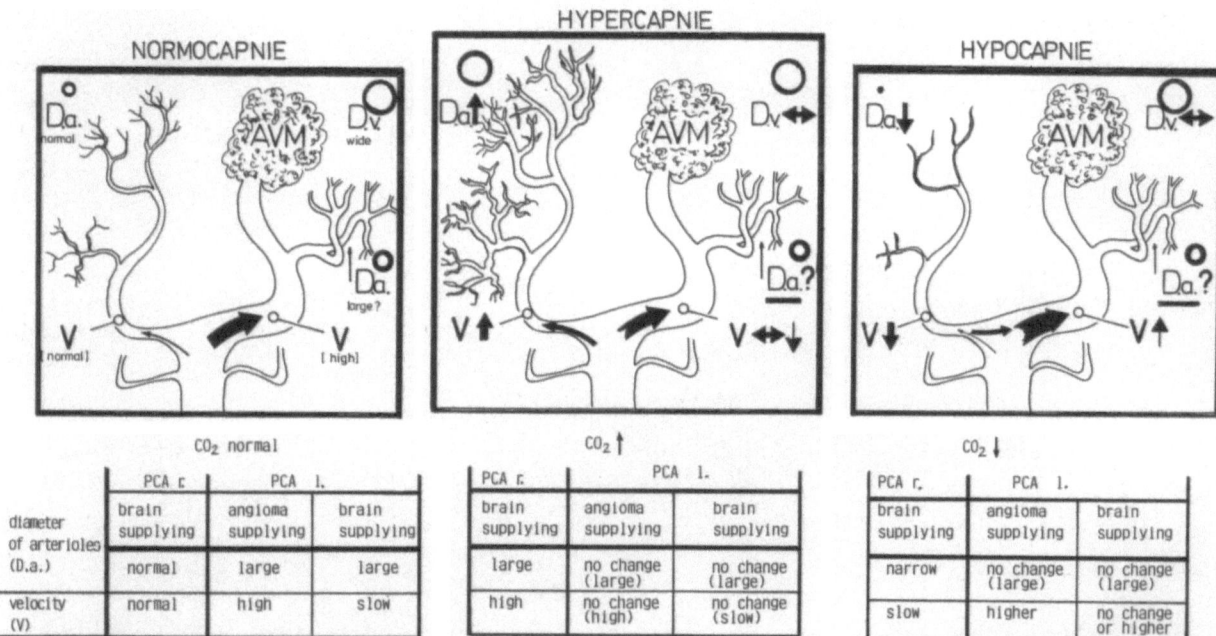

Fig. 120. Intracerebral steal in angiomas induced by pCO_2 changes: In hypocapnia, blood is diverted from normal brain to the AVM; the opposite happens in hypercapnia

It must be stressed that we never had any evidence for postoperative cerebral vasodilation, vasoparalysis and/or real hyperemia even in very large angiomas following angioma surgery. We therefore do not consider that the normal perfusion pressure breakthrough theory provides a convincing explanation for postoperative complications such as hemorrhages. Rather, they would be due to ruptures of the coagulated vessel stumps in the operation site, which are known to have very weak, thin walls without normal muscle layers.

Another interesting finding was that CO_2 reactivity in vasospastic brain arteries was normal in the case we presented. Similar findings have been reported; only very severe vasospasm seemed to result in impaired CO_2 responses (du Boulay 1970, angiography; Heilbrun 1972 and Stattin 1973, CBF measurements).

V. Intraoperative Studies in Angioma Patients

1. Intraoperative Methods (History)

Different authors reported on measurements of perfusion, flow rate, flow velocity and intravasal pressure during brain operations. Application of methods, however, was often complex and time-consuming.

Brain perfusion was measured intraoperatively with radioactive inert gases (Lassen and Ingvar 1961) in

a) patients undergoing therapeutic carotid occlusion because of giant aneurysms (Miller 1977, Gelber 1980, Peerless 1982),

b) extra-intracranial bypass surgery (Cater 1982, Gelber 1980, Little 1980),

c) angioma surgery (Feindel 1971) and

d) aneurysm operations (Griffiths 1974, Merory 1979, Farrar 1981 and Lovick 1982).

Using this indirect method, global and regional changes of brain perfusion can be detected.

Brain perfusion can also be measured by heat emission. Thermal probes have been used by Betz (1963), Wüllenweber (1965, 1968) and Capon (1969). They have also been applied in aneurysm operations (Carter 1982), bypass surgery (Carter 1982), and angioma patients (Carter 1978). This method requires total shielding from external temperature changes and is too sophisticated for intraoperative use.

Fluorescein angiography was applied in bypass operations (Feindel 1967) and angioma surgery (Feindel 1971, Little 1980). Results were poor considering its complicated application, so that it has not been generally adopted.

Electronic auscultation (Bruster 1978, Wintermantel 1982) was used to detect intravasal turbulencies (Ferguson 1972). Flow rate and flow direction are not recorded with this method.

Intraoperative angiography was applied in operations on cavernous sinus fistula (Parkinson 1973), aneurysms (Loop 1966, Wilkins 1972) and angiomas (Loop 1966, Parkinson 1969). It did not gain general acceptance because of limited intraoperative practicability.

The rate of blood flow in single vessels can be recorded electromagnetically (Cannon 1960, Hardesty 1960, Wetterer 1937). This has become a routine method mainly in heart surgery. Intracranial measurements were performed in aneurysm operations (Nornes 1977) and in bypass surgery to detect the flow rate of the feeders (Chater 1976, Crowell 1976, Samson 1978, Spetzler 1980 and Stephens 1980). The probes can only be applied when the vessel's complete circumference is accessible. Correct information on local flow rate and direction of flow is obtained. However, as different vessel diameters require different measuring probes, this procedure is very time-consuming when several vessels are to be examined.

Intraoperative Doppler sonography is easier to handle. It records flow velocity and flow direction. The Doppler method had its breakthrough in noninvasive clinical diagnostics of the cervical arteries (von Reutern 1976, Kreissmann 1978, Büdingen 1982, Renemann 1982, Spencer 1981). The first surgical applications were transcutaneous recordings in operations on cavernous sinus fistula (Matjasko 1975, Büdingen 1978) and intraoperative measurements in subcutaneous angiomas (Bingham 1970, Stephenson 1971). The first recordings in brain surgery were made in angiomas (Hitchon 1979, Nornes 1979 and 1980, Friedrich 1980), in aneurysms (Nornes 1980, Gilsbach 1983) and bypass operations (Gilsbach 1983). High transmitter frequencies and pulsed performance Doppler devices facilitate measurements in small vessels.

Intraarterial measurements of blood pressure were made mainly in the cervical carotid to record the pressure prior to occlusion of this vessel in patients with giant basal aneurysms (Brambilla 1982, Miller 1977, Spetzler 1980). Measurements in cortical arteries have been reported in bypass surgery (Collice 1980, Fuentes 1979, Ito 1980, Yonekawa 1976), in aneurysm operation (Ferguson 1972, Wright 1968) and in angioma surgery (Nornes 1980).

2. Topics of Our Intraoperative Measurements

We studied the intraoperative practicability of Doppler sonography, of electromagnetic flow measurements and intravasal pressure recordings before and after AVM removal. We thereby tried to obtain information on hemodynamics in the angioma as well as in the surrounding brain vessels. Potential implications for the surgical strategy were taken into consideration.

Intraoperative Doppler sonography was carried out in 30 angioma patients, the last 7 of which also underwent electromagnetic and pressure recordings.

Our surgical strategy was to preserve feeding and draining vessels until the AVM had been completely dissected from the surrounding tissue. We assumed that hemostasis should be good as long as perfusion pressure was kept low. The main feeding arteries should therefore be occluded only after dissection of the AVM. This strategy requires exact localization of the malformation and identification of feeding arteries and draining veins which are hard to identify merely by their external appearance.

Intravascular pressure before and after AVM exclusion was recorded. Hemodynamic assessment of angiomas was attempted. In addition, pCO_2-dependent changes of flow velocities in AVM supplying and surrounding vessels were studied. Thereby, the vasomotor activity before and after surgery was tested and Spetzler's theory of suspended autoregulation of cerebral vessels in angioma patients reconsidered.

3. Methods Used in This Study

3.1. Microvascular Intraoperative Doppler Sonography

These measurements were carried out with a device developed by Cathignol (Cathignol 1978, 1980, 1983), which is now commercially available as the MF 20 Microvascular Doppler (EME, Überlingen, FRG). Former intraoperative measurements had been carried out with the same device in aneurysm and bypass surgery (Gilsbach 1983) (Fig. 21).

Technical data:

— pulsed ultrasonic Doppler velocity meter (MF 20),
— transmitted frequency: 20 MHz,
— pulse durations of 250, 450, 850, 1500 ns corresponding to axial gate width of approximately 0.4, 0.7, 1.3, and 2.3 mm (axial resolution),
— lateral gate width (lateral resolution): 1.1 mm with the 3 mm probe,
— automatic gate shifting in 0.1 mm steps with a velocity of 9.25 mm/s,
— pulse repetiton frequencies of 100, 50 and 25 kHz, corresponding to measuring ranges of 7.5, 15, and 30 mm,

— maximum detectable Doppler frequency 12.5 kHz, minimum 0.1 kHz,
— mean frequency by built-in zero-crosser, integration time 0.1–4.0 s,
— calibration by a frequency generator with fixed 1 kHz Doppler frequency,
— direction established by reversal of the mean frequency curves,
— analog output for registration of the mean frequency, audio output for frequency analysis,
— sterilizable miniature probes, 1 cm long, 2 and 3 mm diameter,
— frequency spectrum analysis in real time with FFT.

Intraoperative application

The Doppler shift is angle-dependent. The best signals are obtained in 40–60° angles between probe and vessel (Gilsbach 1983). The smallest vessels investigated were 0.5–1 mm in diameter which is also the diameter of the emitted ultrasonic beam. In such vessels, frequency spectra of the vessel's whole cross-sectional plane are received, whereas in larger vessels the central flow is recorded. Application of the probes is easy. Special preparation of vessels or different probes is not necessary.

3.2. Electromagnetic Measurements of Flow Rate

Electromagnetic measurements were done with the "Narcomatic 500" device (Hugo Sachs Electronic). The probes were 0.7–1.2 mm in diameter. Until now, intracranial measurements had only been performed on branches of the Circle of Willis during aneurysm operations (Nornes 1977). Because of its limited intraoperative practicability, we used this method in only three patients.

Technical data:

Zero-adjustment of the device is maintained automatically throughout the recording. Phasic and mean flow rates are displayed separately. The recording range is 0–200 ml/min; direction of flow is automatically indicated.

Intraoperative application:

After the device has warmed up (15 min), the dampened probes are placed around a completely dissected vessel. Length of dissection has to be 5 mm. The cross-sectional plane of the probe has to be slightly smaller than that of the vessel to ensure contact between electrodes and vessel wall.

3.3. Intravasal Pressure Measurements

Arterial blood pressure in angioma feeders and radial artery was detected with a Statham B 23 ID-transducer and displayed on a Hellige ICU device. Systolic, diastolic and mean values were recorded. Blood pressure curves were displayed together with the electromagnetic recording on a four-channel screen.

Intraoperative application:

The feeding arteries are punctured before and after AVM removal with a very thin needle (0.45 × 13 mm, Braun Melsungen). Measurements of the intravasal blood pressure did not significantly disturb local blood flow because the diameter of the needle was less than half that of the vessel. Following the measurements the needle was removed, the vessel compressed or the puncture site coagulated bipolarly with low current.

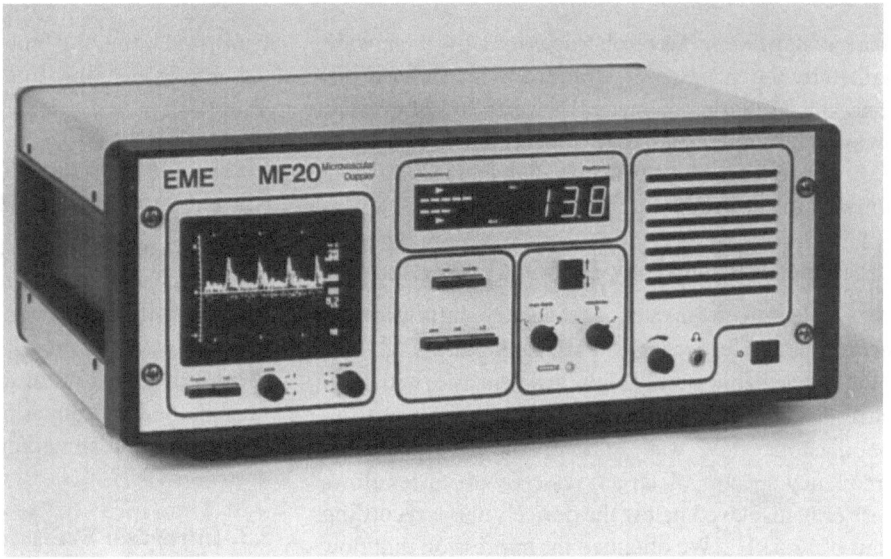

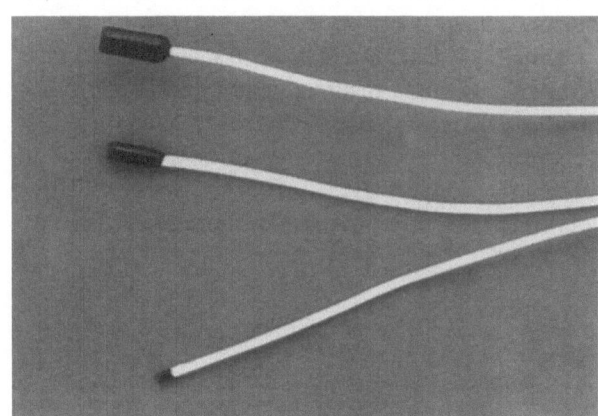

Fig. 121. MF 20-Microvascular Doppler with built-in FFT-Spectrum analyzer with standard miniaturized (1 mm, 2 mm, 3 mm) autoclavable probes (EME, D-7770 Ueberlingen, FRG)

4. Operative Strategy and Preparation of Vessels to be Measured

Superficial angiomas are immediately visible after opening of the dura. In most cases, feeding arteries and arterialized veins cannot be identified by their mere appearance. Following the dopplersonographic measurements the AVMs are dissected as far as possible from cerebral sulci or fissures. We always try to perform the dissection without occluding one of the main feeding or draining vessels. The main purpose of this approach is to keep perfusion pressure of adjacent brain tissue low until the angioma nidus is exposed and thereby prevent bleeding from the neighboring regions.

After the dissection, measurements were performed on the feeders which were temporarily clipped and then reopened for this procedure. The AVM feeders were then excluded in a step-wise fashion to allow time for the brain vessels to adjust to increasing intravascular pressure.

5. Measurements Before AVM Removal

Most of the angiomas were situated within cerebral sulci or fissures and could therefore be exposed without brain tissue resection. Yellowish colorations indicated former occult hemorrhages. Often the arachnoidea was thickened and pale, covering the AVM and its immediate surroundings. The adjacent gyri appeared normal, not atrophic. The number and diameter of brain supplying arteries also seemed normal. Venous drainage was impaired, with dark blood being congested and forced into collateral outflows. In the immediate vicinity, arteries and veins of the AVM could not be identified by their outward appearance.

More distant veins were characterized by their wide diameters. Often however, draining veins showed peripherally decreasing diameters. Nevertheless, they were always much larger than the normal cerebral veins. The feeding arteries showed constant diameters along their course into the AVM.

5.1. Doppler Sonography Before AVM Removal

The whole range of occurring frequency shifts was only recordable in feeders of small angiomas (Fig. 122). They showed abnormally high flow velocities with high diastolic values. The average resistance index was 0.26, the diastolic flow was 74% of the systolic value. Frequency spectra of larger vessels with faster flows were only displayed below the device's upper recording limit of 12.5 kHz. We obtained the impression that flow velocities in smaller vessels were lower than in larger ones. Vessel wall vibrations or turbulencies were not encountered.

Draining veins showed high pulsatile flow velocities in the vicinity of the angioma. Systolic peaks of venous flow were flatter and more rounded off than in arteries. Venous velocities were nevertheless lower than those of arterial feeders and leveled off in the vessel's further course. However, respiratory-dependent deflections as in normal veins did not occur. In hypotension, flow velocity in feeding and draining vessels decreased (Fig. 123).

5.2. Electromagnetic Measurements of Flow Rate

Electromagnetic measurements were carried out in three patients. Depending on the different vessel diameters, different flow rates were found. A 1 mm thick vessel showed 24 ml/min flow, a 1.5 mm vessel had 50 ml/min and flow in a 2 mm vessel was 85 ml/min. This method was abandoned because of the very time-consuming and sometimes risky preparations.

5.3. Intravasal Pressure Measurements

In three patients, intravasal pressure of AVM feeding MCA branches was measured close to the AVM and found to be 59% of systemic mean arterial BP (56%, 59%, 62%). Normally, the mean BP in the MCA is 10–20% below the systemic mean BP, depending on the point of measurement. Therefore, MCA pressure in angioma patients is 20–30% lower than under normal conditions. The MCA pressure pulsations were lower than those of systemic BP (Figs. 124, 125).

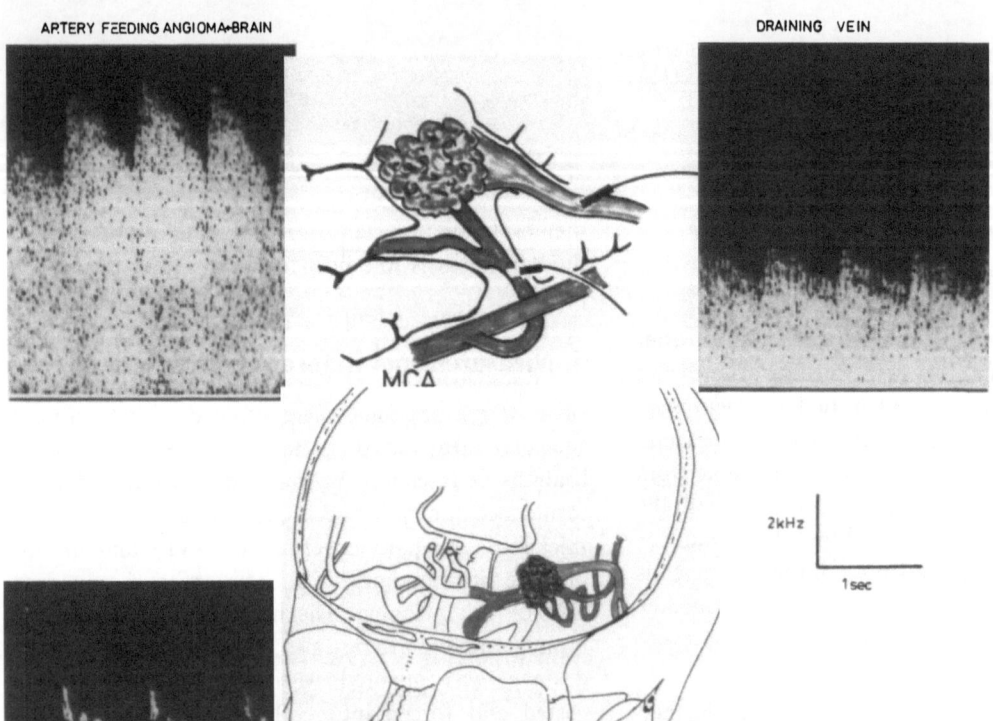

ARTERY FEEDING ANGIOMA+BRAIN

DRAINING VEIN

MCA

2kHz

1sec

Fig. 122. Intraoperative Doppler recordings in a small left frontolateral angioma: The flow velocity is high in the feeding MCA branch (small diameter) and relatively low in the draining vein (large diameter). The vein shows pulsatile flow

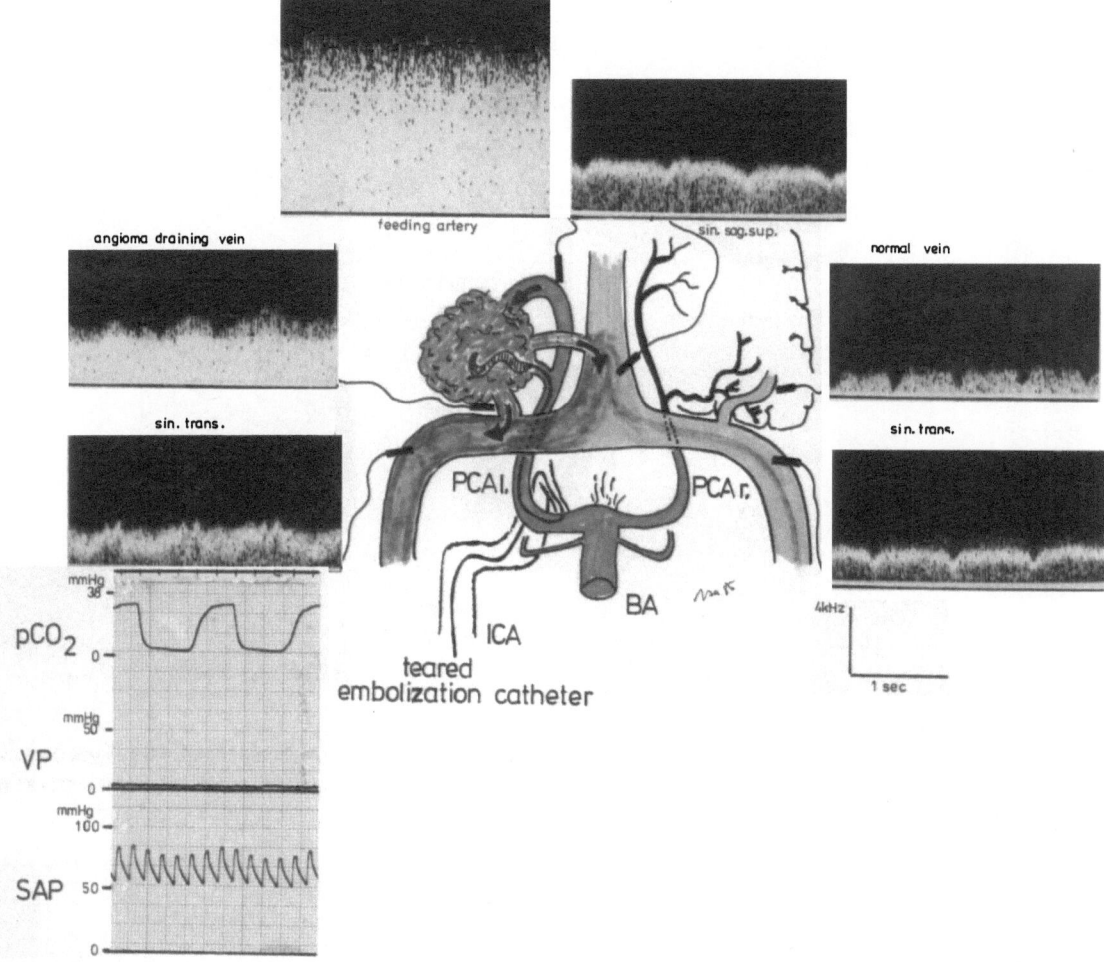

Fig. 123. Intraoperative Doppler recordings and pressure measurements in a left occipital angioma with the patient in a half-seated position: In spite of PEEP ventilation, the intravasal pressure is about zero in the transverse sinus. Flow velocities are recorded from the feeding PCA branch as well as the superior sagittal sinus and both transverse sinuses. The main drainage is via the left transverse sinus and becomes easily detectable. In contrast, normal veins show normal patterns

6. Measurements After AVM Removal

With advancing AVM exclusion, hemorrhages occurred from vessels in the angioma's immediate vicinity. This was mainly due to ruptures of small, thin-walled pial arteries unable to withstand rising intravascular pressure. Pulsation of the brain increased noticeably following the occlusion of the last feeder. As will be shown in the following, this must be attributed to increasing resistance in the arterioles.

Sometimes a local hyperemia was observed. Previously congested veins then drained normally and large AVM draining vessels collapsed. Former AVM veins that did not also receive blood from normal cerebral veins remained red even after AVM exclusion, but they were no longer taut and could be easily squeezed out.

On the other hand, the diameter of the former AVM feeders did not change.

In most cases, the brain volume had slightly increased by the end of surgery. However, massive swelling was only encountered in one intracerebral hematoma.

6.1. Doppler Sonography After AVM Removal

With advancing exclusion of AVM feeders, flow velocity decreased and thereby became recordable (Figs. 126, 127).

Venous flow velocities were also reduced and showed decreasing pulsations.

Flow velocity in the surrounding brain supplying arteries did not change with AVM exclusion.

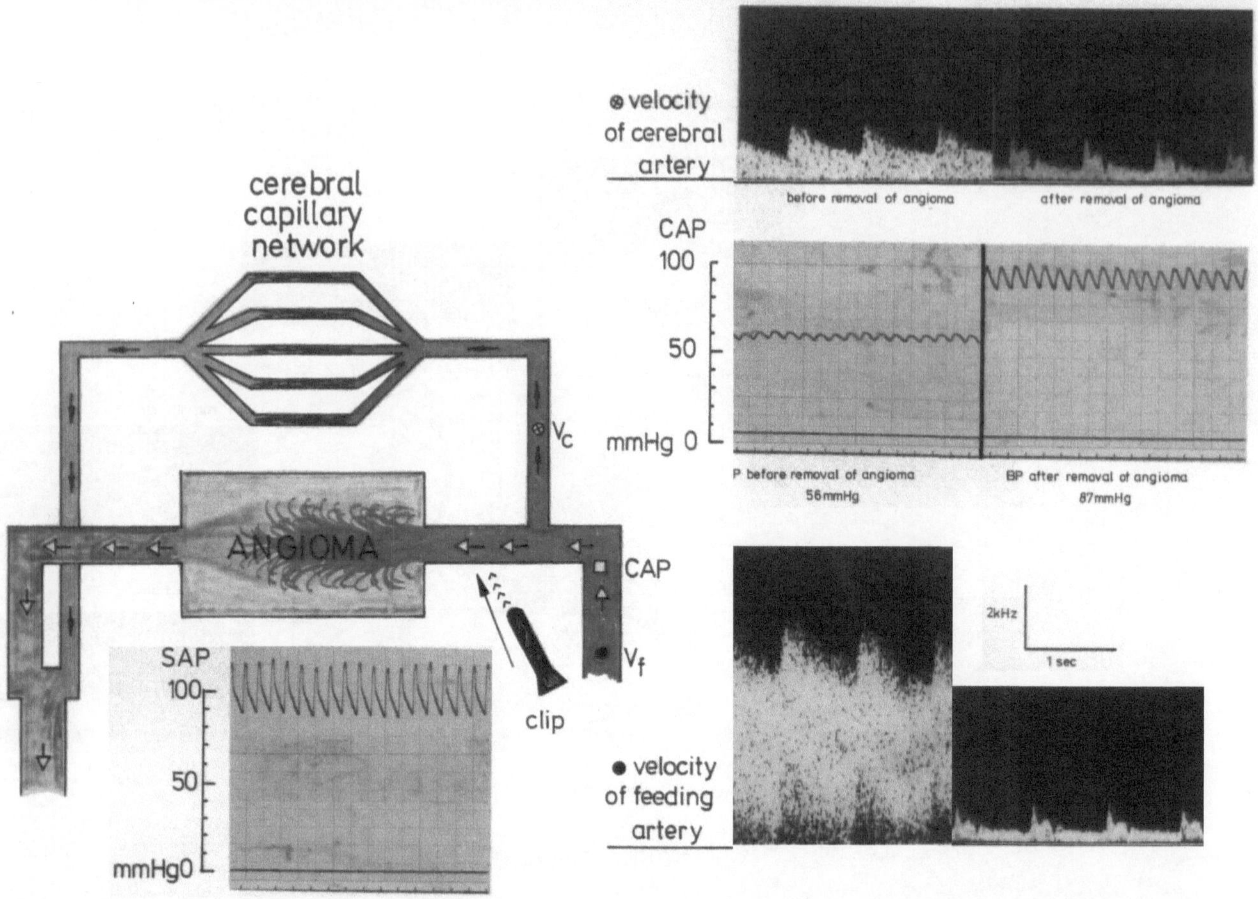

Fig. 124. Schematic illustration of changes in pressure and flow velocity before and after AVM exclusion: After clipping of the feeder, the pressure in cerebral arteries (*CAP*) rises and pressure pulsations increase. Flow velocity in the feeder (*vf*) drops. Velocity in cerebral arteries (*vc*) decreases as well, their patterns indicate raised peripheral stream resistance

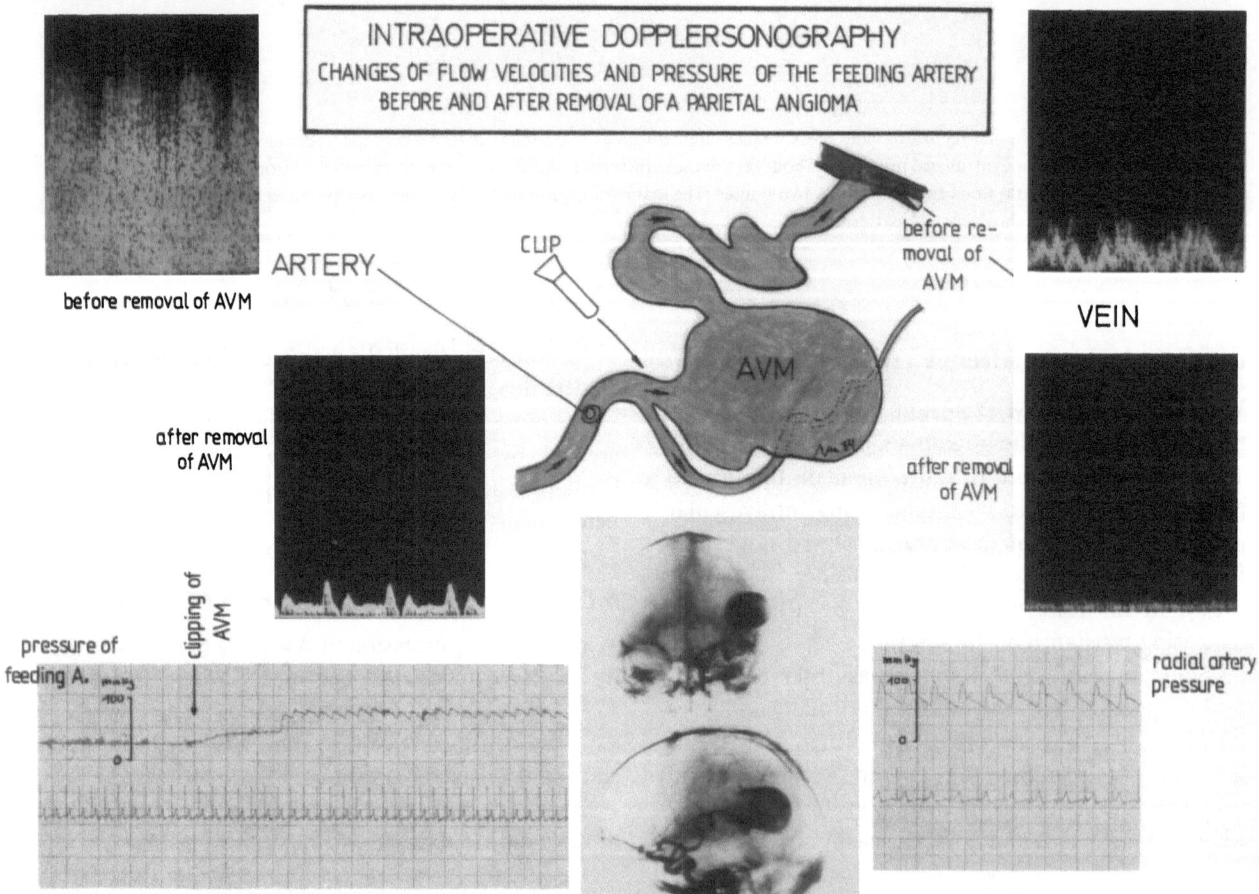

Fig. 125. Intraoperative measurements of flow velocity and pressure in feeding arteries and draining veins of a large AV fistula: Flow velocities

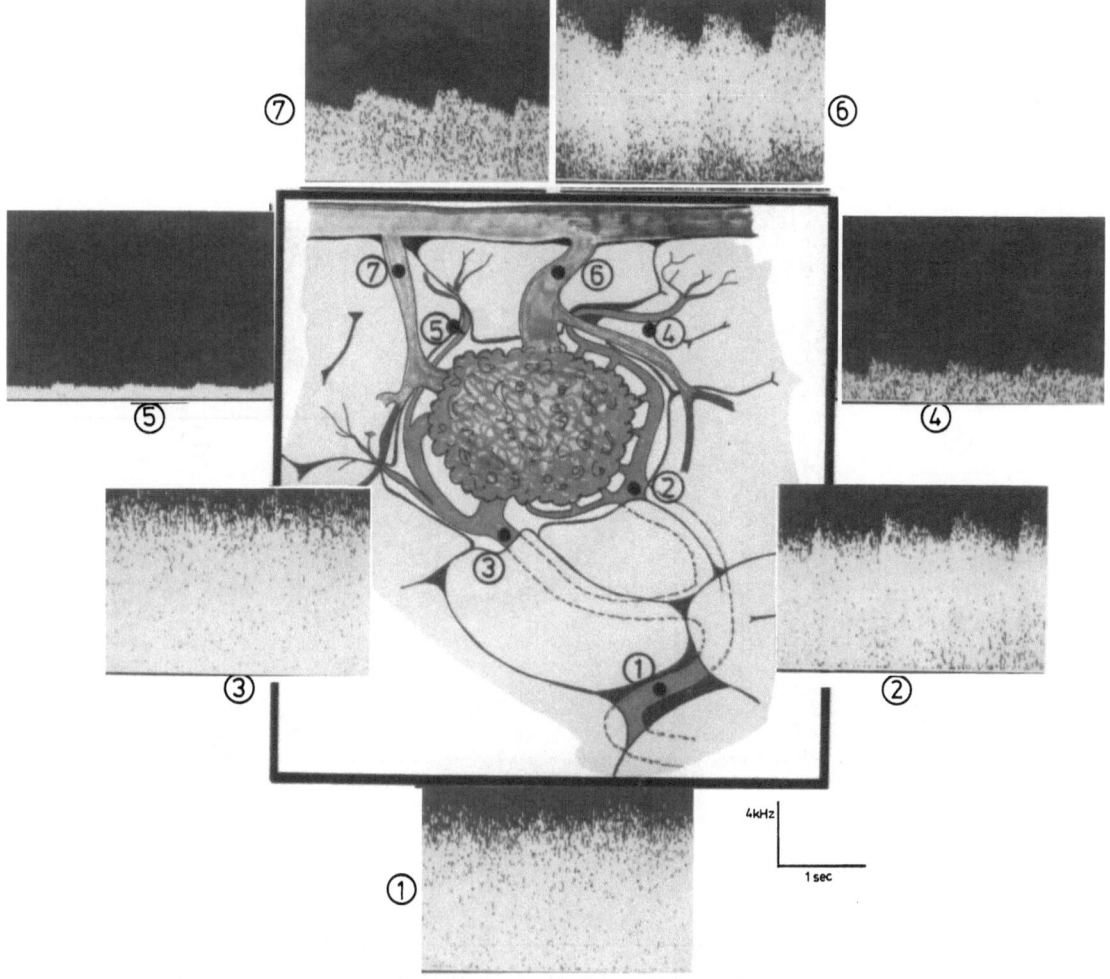

Fig. 126. Intraoperative Doppler recording in the case of a medium-sized parietal angioma: *1* main trunk of the feeder, *2* and *3* feeding arterial branches, *4* and *5* brain-supplying arteries issuing from feeders, *6* and *7* draining veins

Following the angioma exclusion, flow velocity in the main AVM feeding vessels decreased noticeably (Figs. 124, 125, 129).

Frequency spectra were characterized by small systolic peaks and extremely low diastolic signals. These arteries maintained their previous diameters but now had very slow flow. They represent so-called "stagnating arteries" that have been described angiographically (Fig. 130).

Flow patterns in brain supplying branches of the former AVM feeders are also characterized by signs of increased vascular resistance. Systolic flow shows steeply increasing peaks whereas diastolic flow is low. This indicates high peripheral vascular resistance.

Similar flow patterns even occur in surrounding brain arteries which did not contribute to the angioma. The increase in vascular resistance however, is less significant than in the aforementioned vessels.

Flow velocities in the former AVM draining veins often decrease below the recording range. Frequency spectra then show flat, respiratory-dependent deflections (Fig. 131).

6.2. Intravasal Pressure Measurements After AVM Removal

Arterial BP in former AVM feeders increased by an average of 53.5% (Table 23). This increase did not depend on the angioma's size nor on the length or diameter of its feeders. Following AVM exclusion, local BP was 93% that of the systemic BP (91%, 92%, 96%). The diameters of previously AVM supplying arteries did not change. This corresponds to postoperative angiographic studies where stagnating arteries maintain their preoperative diameters or may even slightly enlarge.

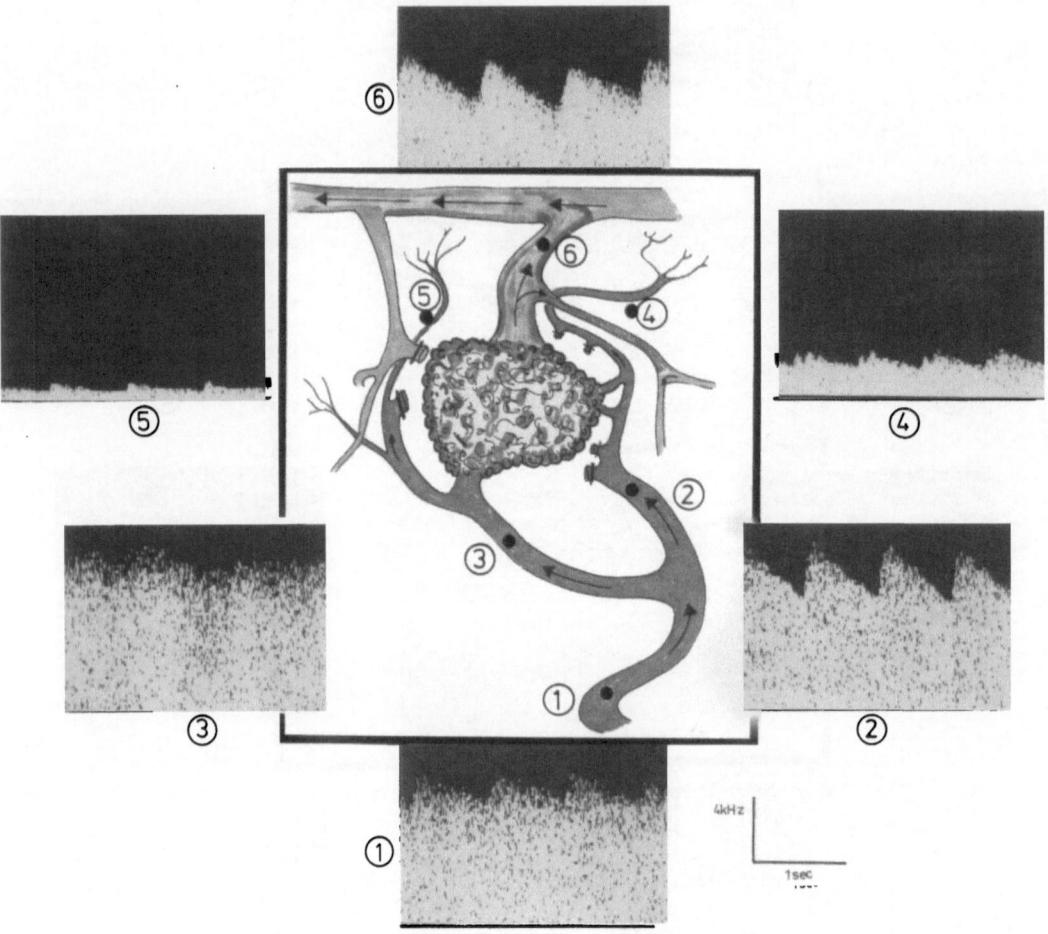

Fig. 127. Intraoperative Doppler recording after partial exclusion of the angioma (same case as Fig. 126): Flow velocity in vessel 2 has decreased and now lies within the recording range showing pulsatile flow

Table 23. *Intravascular Pressure (P) Before and After AVM Removal, n = 3*

Patient	P in radial artery (mmHg)			Systolic-diastolic mean P in angioma feeder before removal (mmHg)			Sysolic-diastolic mean P in angioma feeder after removal (mmHg)			Change of mean P before and after removal (%)	Length and diameter of angioma feeder
	syst.	diast.	MP	syst.	diast.	MP	syst.	diast.	MP		
N. PH.	100	59	72.6	50	39.3	44.0	80.3	60	66.7	51.6	16.0 cm 2.7 mm
H. K.	119.7	89.5	95.5	56	52	53.5	96	83	87.3	53.2	12.0 cm 3.5 mm
G. H.	100	64	76			47.3	84.2	68.4	73.6	55.6	8.0 cm 4.0 mm

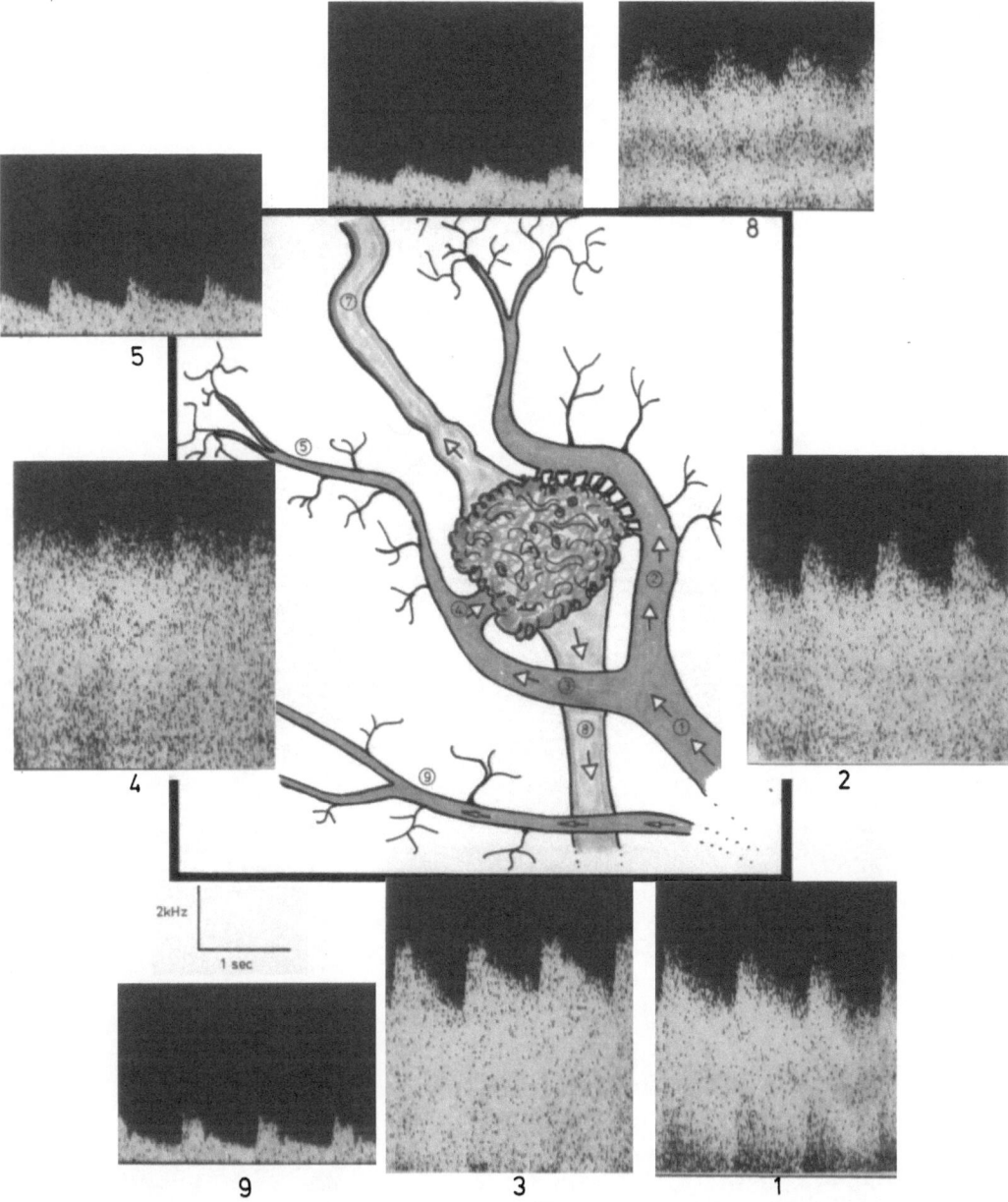

Fig. 128. Intraoperative Doppler recordings in the case of a parietal angioma: Flow velocities in feeding arteries (*1, 2*), in cerebral arteries issuing from them (*5*), in draining veins (*7*) as well as a neighboring normal brain artery (*9*)

Thus, preoperative fast flow and low pressure AVM feeders postoperatively become very low flow and high pressure vessels with unchanged diameters.

6.3. Electromagnetic Measurements

Flow rates after AVM removal were below the recording range of our device.

7. CO₂ Reactivity of Vessels Supplying Angioma and Brain Before and After AVM Exclusion

In five patients, intraoperative Doppler measurements were performed with varying pCO₂ values before and after AVM removal. Under hypercapnic conditions the brain was found to enlarge. It returned to its previous size when pCO₂ was reduced. Systemic BP remained unchanged.

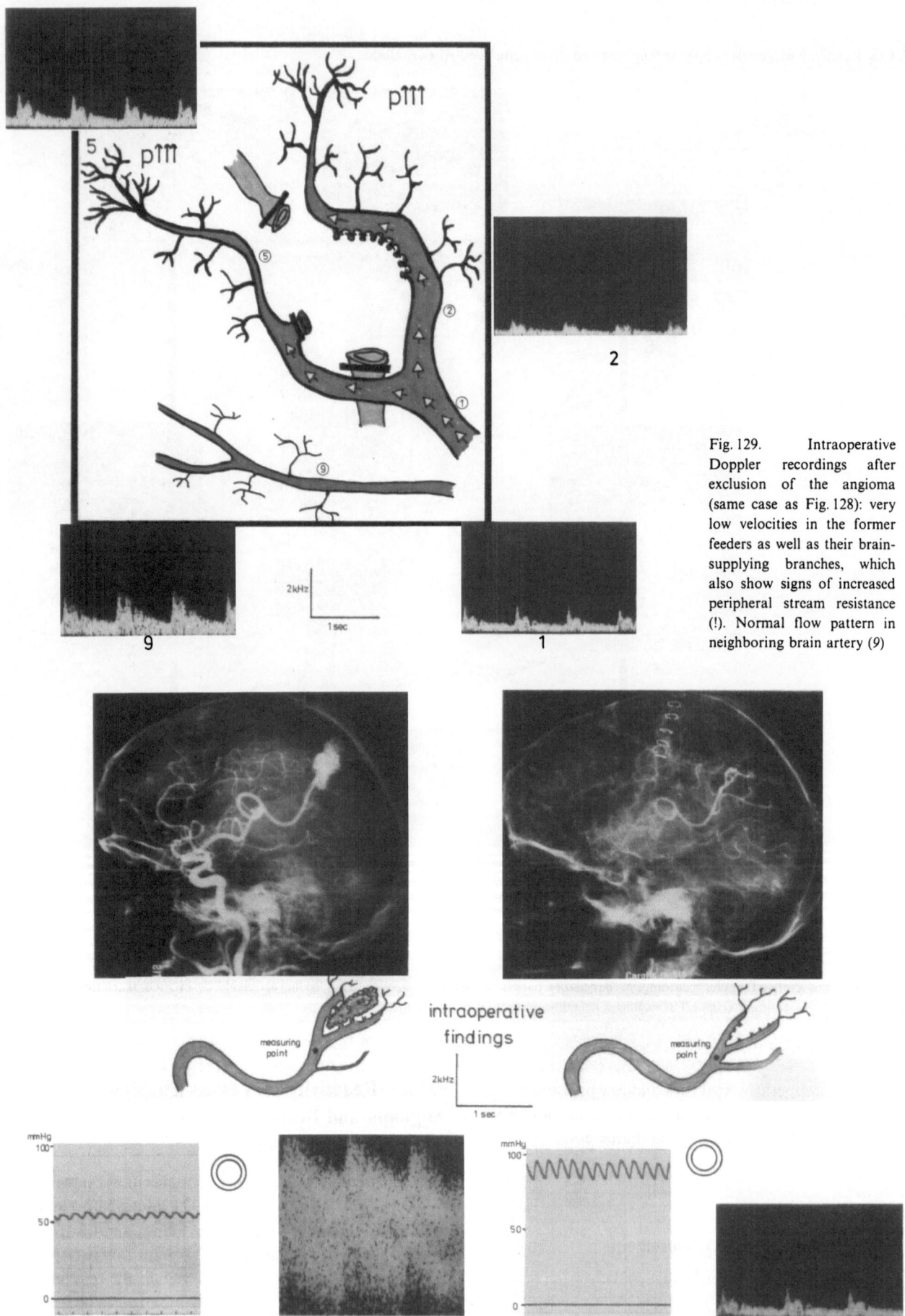

Fig. 129. Intraoperative Doppler recordings after exclusion of the angioma (same case as Fig. 128): very low velocities in the former feeders as well as their brain-supplying branches, which also show signs of increased peripheral stream resistance (!). Normal flow pattern in neighboring brain artery (9)

intraoperative findings

Fig. 130. Angiographic, Doppler and pressure findings in a former main feeding artery (right MCA) after AVM removal: very slow flow, high intravasal pressure and unchanged diameter ("stagnating artery")

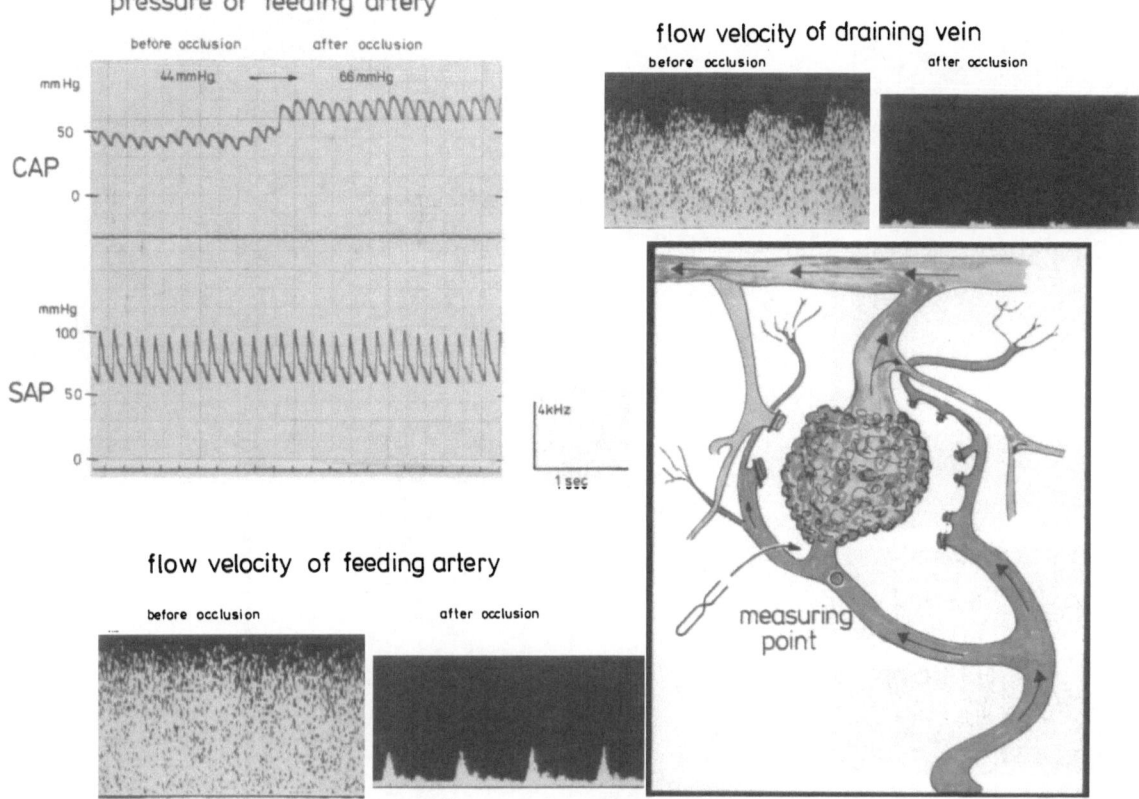

Fig. 131. Doppler and pressure recordings before and after exclusion of a medium-sized right parietal angioma: Clipping of the last feeder results in a local arterial pressure increase by 51.6%; the previously diverted venous flow from normal brain now runs directly into the superior sagittal sinus

7.1. CO$_2$ Reactivity Before AVM Removal

In hypercapnia, flow velocity in AVM supplying arteries only increased when the vessel concerned was also brain supplying. The extent of this feeder's reaction was found to be directly proportional to the volume of supplied brain. Flow velocities did not change significantly under hypocapnia (Fig. 132). In general, the vasomotor response was poor.

Brain arteries branching from the feeder distal of the AVM show high diastolic flow velocities even under normocapnia. With increased pCO$_2$, systolic and diastolic values climbed further. Under hypocapnic conditions, only the systolic velocity became slightly reduced, whereas the diastolic velocity maintained its level. This poor diastolic response may be due to chronic dilation of the brain arterioles surrounding the AVM. It also demonstrates that the autoregulative force of intravascular hypotension prevails over that of hypocapnia.

Brain arteries that did not contribute to the AVM showed normal vasomotor responses to pCO$_2$ changes.

7.2. CO$_2$ Reactivity After AVM Removal

Following angioma exclusion, flow velocities were very low in the main feeding vessels.

Brain supplying arteries branching from these feeders distal of the former AVM also showed decreased velocities.

Under hypercapnic conditions two-fold flow velocities were often found in all vessels involved. Hypocapnia resulted in a considerable drop in flow velocity, so that it was often hardly measurable (Fig. 133). An impairment of vasomotor response to pCO$_2$ changes was therefore not found after AVM removal. On the contrary, vasomotricity almost overshot.

These findings do not accord with the breakthrough conception. They are considered quite essential and should be illustrated with some examples:

Patient 1: Small frontolateral angioma supplied by MCA branch (Fig. 134)

The MCA branch supplying angioma and brain showed some response to pCO$_2$ changes before AVM removal. After surgery this

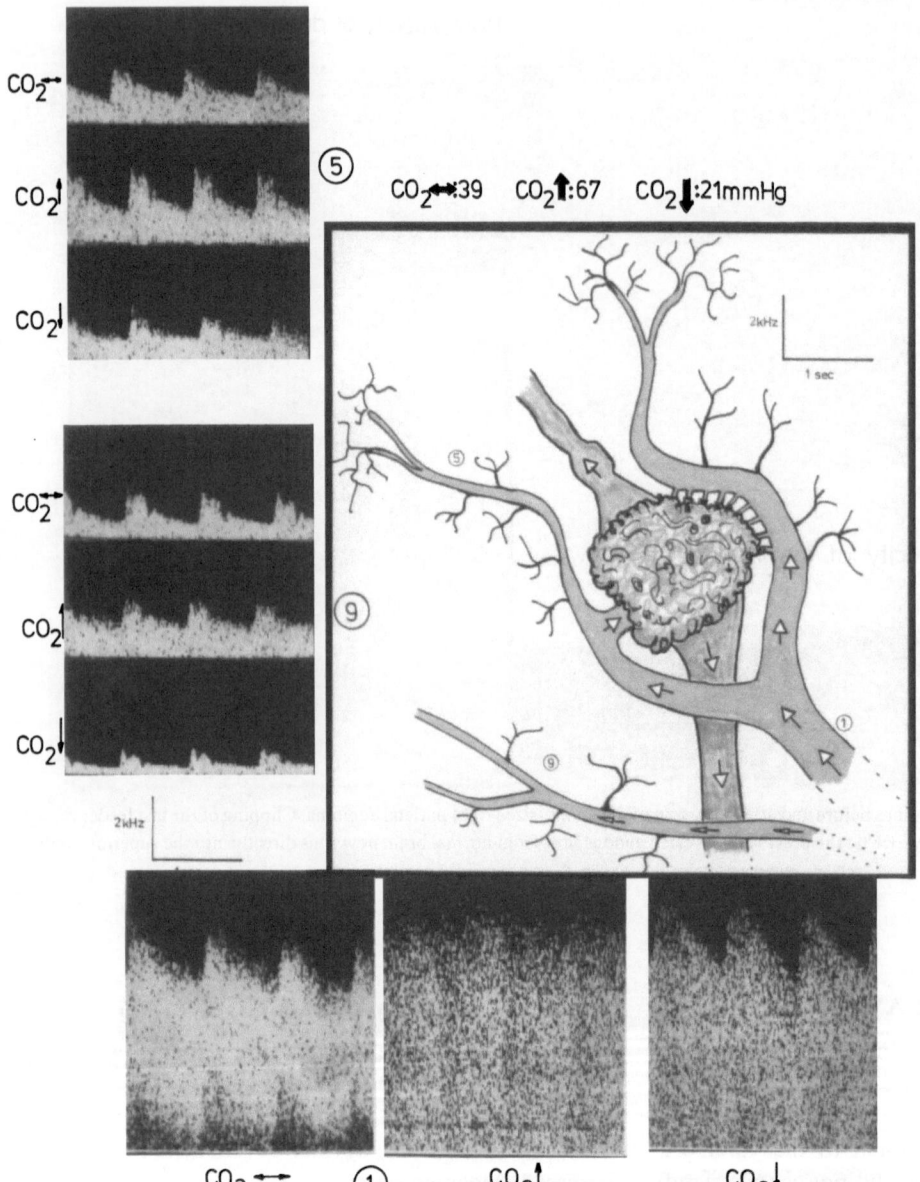

Fig. 132. Intraoperative Doppler measurements of the CO_2 reactivity in AVM feeders (*1*), brain supplying arteries issuing from them (*5*) and a neighboring normal brain artery (*9*)

branch had the typical flow pattern of a stagnating artery with very low velocities. In spite of high vascular resistance, this vessel responded to pCO_2 changes with increasing velocity under hypercapnic and decreasing velocity under hypocapnic conditions. The neighboring MCA branch that did not contribute to the angioma showed completely normal pCO_2 response.

Patient 2: Medium-sized right temporobasal AVM with main supply from PCA (Fig. 135).

After AVM removal, flow velocity of a brain supplying branch of the feeder showed high systolic peaks and very low diastolic values, indicating high vascular resistance. Elevation of pCO_2 up to 70 mmHg resulted in significantly increasing systolic and diastolic velocities, whereas lowered pCO_2 (21 mmHg) led to a further velocity decrease with undetectable enddiastolic values.

Patient 3: Medium-sized parietal angioma supplied by MCA, ACA and PCA (Fig. 136).

In this case, CO_2 reactivity was studied after exclusion of the supply from ACA, PCA and also parts of the MCA. Branch 2, supplying AVM and brain, already reacted to hypocapnia. Exclusively AVM feeding branch 3 did not respond. Eventual increases in velocity in branch 2 under hypercapnia were not recordable as the recording range had been previously exceeded.

Brain supplying branches leaving the feeder distal from the AVM still did not show diastolic flow reduction in hypocapnia. Reactivity to hypercapnia was normal. This demonstrates that peripheral vascular resistance does not increase until all AVM feeders are excluded.

Venous flow did not react to pCO_2 changes.

In addition to the Doppler recordings, these three patients also

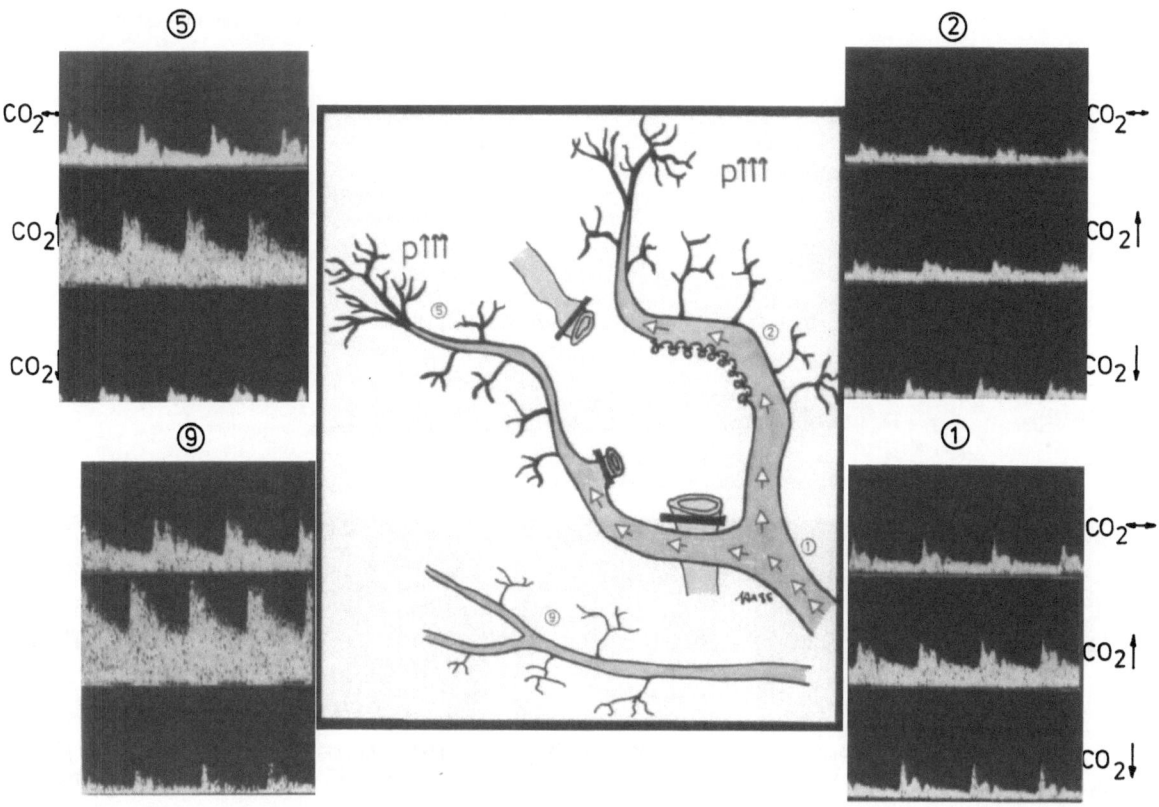

Fig. 133. Intraoperative Doppler measurements of the CO_2 reactivity after AVM exclusion (same case as Fig. 127): normal responses to pCO_2 changes

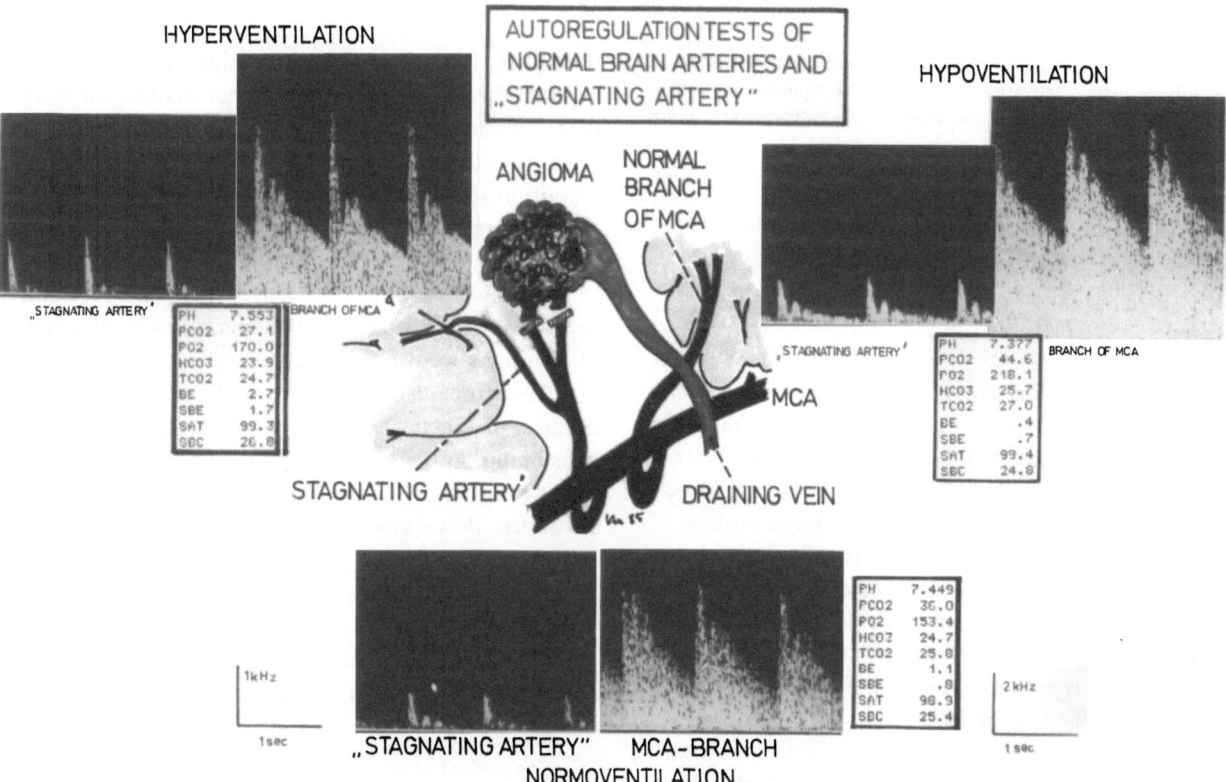

Fig. 134. Intraoperative Doppler measurements of the CO_2 reactivity after exclusion of a small angioma: normal CO_2 response of the normal MCA branch. The brain-supplying branch issuing from the AVM feeder ("stagnating artery") shows signs of very high stream resistance with the CO_2 reactivity still preserved

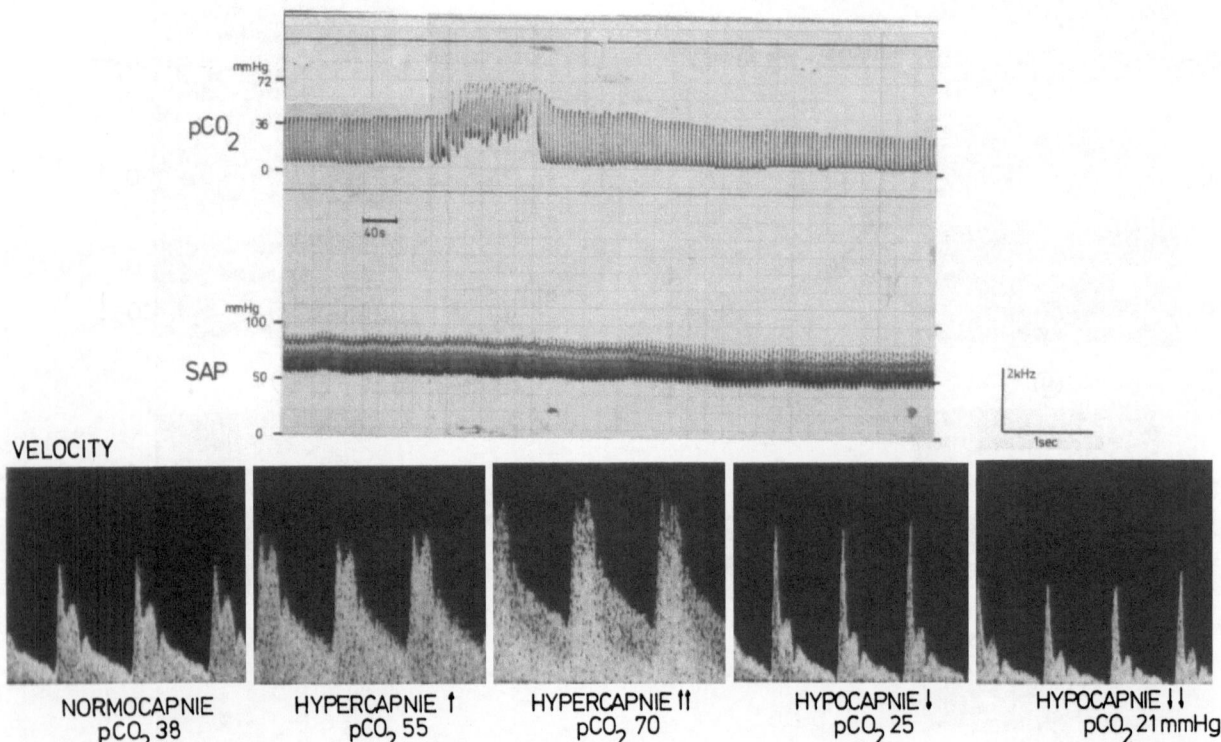

Fig. 135. CO_2 reactivity after removal of a left temporobasal angioma; simultaneous recordings of endexpiratory pCO_2, systemic arterial BP and flow velocity in a brain supplying branch of the former feeder: normal responses to hypercapnia and hypocapnia

underwent electromagnetic measurements and intravasal pressure recordings before and after complete AVM removal. Flow rate dropped from 26 ml/min (1 mm thick vessel) to values below the recording range. Intravasal pressure increased by 51.6% from 44 mmHg to 66 mmHg. High flow velocities that had previously not been recordable showed a decrease just below the upper recording limit. The velocity of venous flow was no longer detectable following the operation (Fig. 137).

8. Discussion

The first dopplersonographic studies on angiomas were performed by Hitchon 1979 and Friedrich 1980. The equipment they used was unsuitable for small vessels, since it was of the continuous wave type with relatively low frequencies of 4 and 5 MHz and large probes.

Nornes was the first to use a pulsed Doppler system for intraoperative studies of angiomas, first with a 6 MHz operating frequency (Nornes 1979) and then with a 10 MHz system (Nornes 1980). Equipment of a higher frequency, however, will make recordings in smaller vessels more precise.

The Doppler prototype device used in our studies was a pulsed system with an ultrasonic beam of 20 MHz (Cathignol 1980). Small measuring probes facilitate reliable intraoperative measurements in small vessels

(Gilsbach 1983). Nevertheless, high flow velocities cannot be recorded because the upper recording limit of the equipment is 12.5 kHz (Cathignol 1983). At present, the use of different filter combinations and higher pulse repetition frequencies is under evaluation in order to refine intraoperative Doppler recording techniques (Alec Eden, EME, Ueberlingen 1985)*.

As flow velocities in feeding arteries were not recordable in most cases, intraoperative classification of angiomas according to velocity criteria was impossible. Electromagnetic measurements were unsuitable because of practical use being complicated and time-consuming. As did Nornes, we found velocities in small feeding vessels to be lower than in larger ones. However, flow velocity was recordable in only one patient.

Arteries and veins of an AVM could be differentiated in some distance from the malformation, but it was difficult and sometimes impossible in its immediate surroundings due to high and pulsatile

---

* As a result of these trials we are since using an improved instrument with a built-in FFT spectrum analyzer and a filtering system which permits measurements of up to 22 kHz with autoclavable probes from 1 mm diameter (EME, Ueberlingen, FRG).

$CO_2 \leftrightarrow$: 38 mmHg

$CO_2 \uparrow$: 65 mmHg

$CO_2 \downarrow$: 20mmHg

intraoperative autoregulation
test after partial skeletisation
of parietal angioma

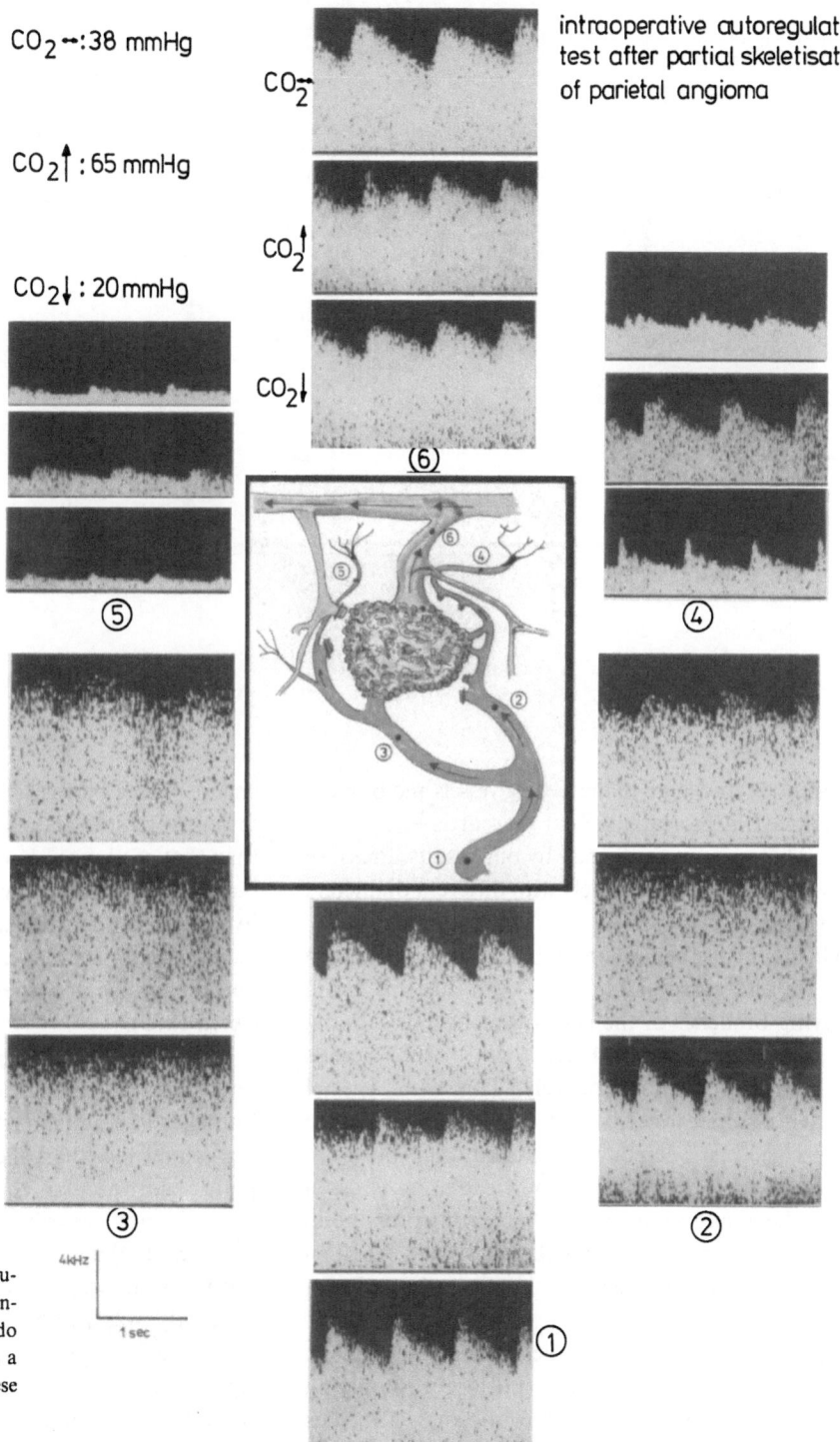

Fig. 136. CO_2 reactivity after partial exclusion of a parietal angioma: brain-supplying branches (4) of the feeders do not respond to hypocapnia indicating a still low intravascular pressure in these vessels

venous flow. However, detection of blood flow direction to or from the AVM was easy.

Complete exclusion of an angioma could be proven. This was often difficult through mere visual inspection because larger veins were still filled with red blood. Venous flow velocities however, were hardly measurable after AVM exclusion (stagnating venous flow).

In accordance with Nornes (1980), intravasal pressure was found to increase after AVM exclusion. The mean increase in our series was 53.5% (Nornes: 50%) and did not depend on the angioma's previous size. This is easy to understand, since measurements were made proximal of the AVM where the highest pressure drop occurs. Nornes considered the pressure drop to be

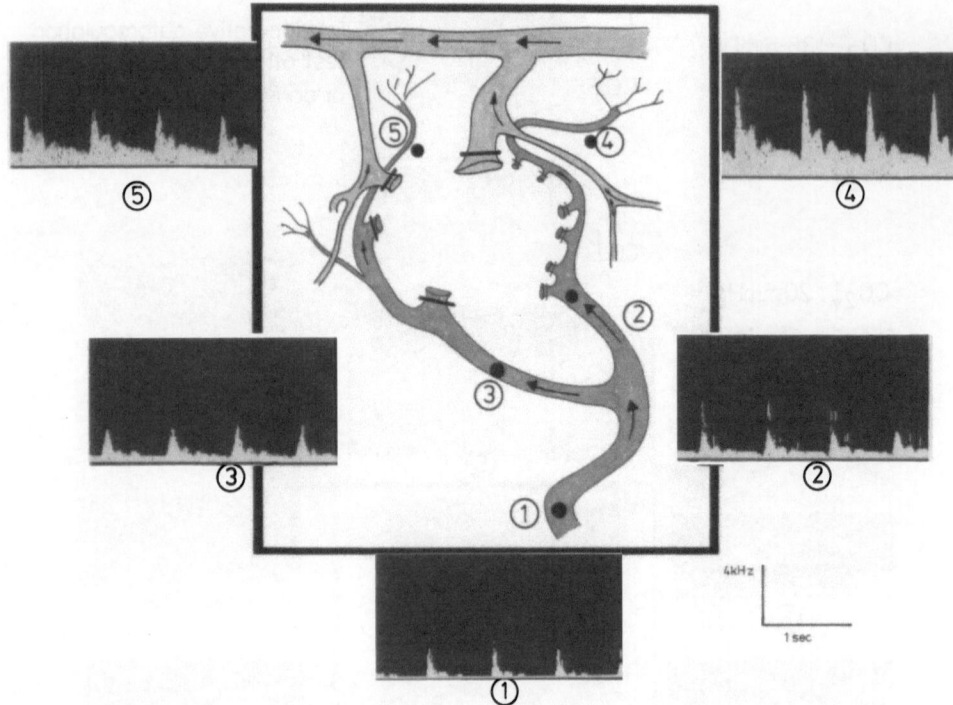

Fig. 137. Intraoperative Doppler sonography in former AVM feeders and their brain-supplying branches following AVM exclusion: all vessels show signs of raised peripheral vascular resistance

dependent on the length of the feeding vessels and be most pronounced in more than 8 cm long arteries (Nornes 1980). This was not confirmed by our results. Pressure drops were the same in vessels with different diameters or lengths.

This report is the first on Doppler measurements in brain vessels surrounding the AVM and in brain supplying branches of the AVM feeders. For the first time, investigations of CO_2 reactivity of such vessels were performed intraoperatively. This is of special interest, as these vessels had been thought to be paralyzed and thus demonstrate breakthrough phenomena (Lassen 1971) following AVM exclusion (Lassen 1971). The assumption was that increasing pressure in vessels with impaired autoregulation should result in overperfusion, hemorrhage and brain swelling (Spetzler 1978, Siesjö 1971).

We found high flow velocities in angioma feeders. Surrounding cerebral arteries and brain supplying branches of AVM feeders show normal velocities though with relatively high diastolic values, indicating lowered peripheral stream resistance in surrounding brain tissue. In these vessels, velocities increase under hypercapnic conditions, whereas hypocapnia leaves the diastolic flow unchanged. This fact must be explained by basic principles of vascular autoregulation. Brain supplying arterioles issuing from the AVM feeder have low intravasal pressure and therefore dilate in order to

ensure brain perfusion. The autoregulative force of lowered pressure in these vessels was obviously stronger than that of a lowered pCO_2. Similar findings have been made in dogs where vascular CO_2 reactivity under hypotension was studied (Harper 1965). Other authors recently reported that in nitroglycerine-induced hypotension no vasomotor response to lowered pCO_2 occurs (Artu 1985).

Following AVM removal, flow velocity decreases noticeably in the feeders and the brain supplying branches issuing from them. They show steep systolic peaks and low diastolic values which' indicate raised peripheral vascular resistance. Surrounding arteries that did not contribute to the angioma showed no significant change. This demonstrates intact vasomotricity in previously dilated arterioles. Also, CO_2 reactivity was established at once. Hypercapnia led to increasing flow velocities. Hypocapnia caused a significant decrease, so that diastolic flow velocities were often no longer recordable.

Our observations were made only on medium-sized angiomas. In very large AVMs, CO_2 reactivity has not yet been studied intraoperatively (but transcranially). As large AVMs show the same intravasal pressures prior to the operation, and postoperative pressure increases cannot be more pronounced than in our cases, no fundamental differences should be expected.

Our results question the basic assumption of the

breakthrough theory. Impaired vasomotricity after long-standing dilation of vessels in the angiomas immediate vicinity is not demonstrable.

9. Summary of Intraoperative Measurements

Intraoperative Doppler sonography is a good support in realizing operative strategies in angioma surgery. After arterial and venous vessels have been identified, dissection can be adjusted to hemodynamic characteristics of the individual case. Electromagnetic flow measurements were not practicable. Intravasal pressure measurements demonstrated a 53.5% increase in feeding arteries following the AVM exclusion. We consider this to be the main cause of complications. Pathologically thin-walled vessels that have been coagulated may not withstand rising intravascular pressure and thus rupture. The so-called local "hyperemia" in the surgical site, as sometimes encountered intraoperatively, represents a vascular congestion in small arteries proximal to the resistance vessels. These arteries sometimes dilate after AVM exclusion. A hemodynamically relevant, real hyperemia in the sense of increased blood flow rates was never demonstrable in our intraoperative investigations during and after angioma removal. Studies of the CO_2 response showed that arterioles surrounding the AVM have excellent vasomotricity and tend to constrict. There was no indication of "normal pressure breakthrough" phenomena.

VI. Final Conclusions

Our animal experiments supplied precise information on hemodynamic characteristics in vessels with high flow velocities and of vessels connected in parallel to AV fistulas. Comparable investigations did not exist until now. It was shown that blood flow rates in cervical AV fistulas increase to about ten times the normal values. This increase is limited only by the cardiac output volume, by the diameters of fistula vessels, by the cross-sectional plane of the shunt and by systemic blood pressure. The stream resistance in the fistula feeding vessel drops with minimal resistance indices reaching 0.28. The flow rates and vascular resistances in experimental cervical AV fistulas were the same as in human AVMs.

Hemodynamic changes in experimental AV fistulas result in predominantly venous vessel wall alterations. Stenoses in long-term fistulas may develop and then result in declining shunt flow rates. Similar venous alterations have been described in dural AVM and may lead to sinus thrombosis.

In the experimental H-fistula model, cortical microcirculation is reduced by 25%. Spetzler's model only slightly affects microcirculation. In neither of these animal models did we have any evidence for a disturbance of vascular autoregulation following the sudden occlusion of chronic fistulas. The blood brain barrier was intact. As under normal conditions, only blood pressure increases above the upper limit of vascular autoregulation results in Evans blue extravasations.

New information was also obtained by our clinical investigations in human AVMs. Transcranial Doppler sonography was found to be most valuable in estimating hemodynamic force of an angioma and the hemodynamic changes after AVM-removal. Transcranial identification of angioma feeders is possible; only small malformations below 2 ccm of volume were not detectable. Nevertheless, angiography remains essential in terms of diagnosis and surgical planning. Transcranial Doppler sonography, however, offers additional information on the hemodynamic significance of an angioma (blood distribution, steal effects). Different feeders can be graduated according to their degree of participation in the AVM supply. This may be of relevance in planning one-step or step-wise removal. In the first operative step, one could try to transform exclusive AVM feeders into partial feeders by cutting down the number of shunts.

Tests of the CO_2 reactivity yielded information on the amount of brain being supplied by AVM feeding arteries. The more impaired the response of such vessels to hypocapnia, the higher is their degree of AVM supply. Reduced CO_2 responses of neighboring brain arteries that do not contribute to the angioma, especially a limited hypocapnic diastolic flow deceleration, indicates hemodynamic involvement of normal brain.

Intraoperative recordings confirmed experimental findings that the currently available Doppler devices do not allow measurements of fast flow velocities in angiomas. Differentiation of venous and arterial vessels was possible, so that early occlusion of major draining veins could be avoided. Following the AVM removal, complete exclusion was proven by Doppler sonography in all accessible areas.

We intraoperatively compared electromagnetic (flow rate) and Doppler sonographic (flow velocity) recording techniques. Electromagnetic measurements were superior in the experimental situation. Intraoperative applications in humans, however, are very time-consuming and therefore sometimes precarious with this method. Doppler recordings are much more practicable. Veins can be easily identified; the hemodynamic situation can be assessed during and after AVM exclusion.

Following the removal of angiomas, intraoperative and transcranial Doppler sonography as well as angiography reveals important hemodynamic changes. Flow velocities in former AVM feeders are considerably reduced with often undetectable diastolic values that indicate a sharp increase in arteriolar resistance ("stagnating" arteries). Accordingly, hyperperfusion or real hyperemia was never demonstrable, even in the former AVM's immediate surroundings. CO_2-dependent vasomotricity was excellent following AVM removal.

These findings contradict general conceptions that are based upon the normal perfusion pressure breakthrough theory.

Angiographic results correspond to the postoperative Doppler findings. "Stagnating" arteries show the same diameters as preoperative and extremely low flow. Such former AVM feeders remain visible even throughout the venous angiographical phase; they remain demonstrable up to 3 weeks following surgery. Intraoperative measurements in these vessels show an increase of 53.5% in intravascular pressure. No angiographic signs of postoperative hyperemia or hyperperfusion are seen; previous impairment of blood distribution and steal phenomena disappear.

Transcranial Doppler measurements after the removal of angiomas confirmed our intraoperative and angiographic findings. In the immediate postoperative course, flow velocities drop significantly and then return to normal in the following days. After that, flow velocities in the pathological hemisphere maintain slightly lower levels than contralateral.

Postoperative transcranial Doppler recordings as well as intraoperative investigation of the CO_2 reactivity mainly reveal an initially reduced diastolic flow deceleration under hypocapnia. This is due to maximal vasoconstriction, which leaves the arterioles unable to react further. This subsides within a maximum of 7 days. The duration of postoperative arteriolar vasoconstriction mainly depends on the angioma's previous size.

Based upon our experimental and clinical investigations, the following conclusions must be drawn:

—Hyperperfusion does not occur after angioma removal because the flow velocities neither increase in former AVM feeders nor in the surrounding brain arteries.

—The "normal perfusion pressure breakthrough" theory does not prove to be correct, because postoperative vasomotricity of the cerebral resistance vessels is excellent.

—"Circulatory breakthrough" (Nornes 1977) is one of the factors causing postoperative complications. Thinwalled vessel stumps in the surgical site may not withstand high intravasal pressures that have increased by 53.5%.

—Complications may also be due to thrombosis. As postoperative blood flow in stagnating arteries and enlarged veins is very low for 8–10 days, thrombotic infarcts of arterial and venous origin may occur and be further complicated by secondary hemorrhages.

References

Aaslid R, Huber P, Nornes H (1984) Evaluation of cerebrovascular spasm with transcranial Doppler ultrasound. J Neurosurg 60: 37–41

— Nornes N (1984) Musical murmurs in human cerebral arteries after subarachnoid hemorrhage. J Neurosurg 60: 32–36

— Markwalder TM, Nornes H (1982) Noninvasive transcranial Doppler ultrasound recording of flow velocity in basal cerebral arteries. J Neurosurg 57: 769–774

Ackerman RH, Zilkha E, Bull JWD, Du Boulay GH, Mashall J, Ross Rusell RW, Symon L (1973) The relationship of the CO_2 reactivity of cerebral vessels to blood pressure and mean resting blood flow. Neurology 23: 21–26

Amundsen AK, Amundsen P, Refsum H (1966) Circulation time and pattern in cerebral angiography using different techniques for general anaesthesia. Acta Radiol (Diagn) 5: 84–90

Ancri D, Pertuiset B (1985) Mesure des vitesses sanguines instantanées dans les artères carotides internes par Doppler pulse dans les malformations artérioveineuses cérébrales. Neurochirurgie 31: 1–6

Arnolds BJ, von Reutern GM (1986) Transcranial Dopplersonography. Examination technique and normal reference values. Ultrasound Med Biol (in print)

Aoki H, Hilgers H, Brown Jr PH D, Kittle CF (1963) Hemodynamic effects of hypercapnia. Surgery 14: 232

Aronson SM (1972) Vascular malformations. In: Minckler J (ed) Pathology of the nervous system. McGraw-Hill, New York

Artu AA (1985) Cerebral vascular response to hypocapnia during nitroglycerin-induced hypotension. Neurosurgery 16: 468–472

Auer LM (1978) The pathogenesis of hypertensive encephalopathy. Acta Neurochir [Suppl] 27. Springer, Wien New York

Axel L (1980) Cerebral blood flow determination by rapid-sequence computed tomography. Radiology 137: 679–686

Bada HS, Korones SB, Magill HL (1982) Alterations in cerebral hemodynamics in relation to onset of neonatal intraventricular hemorrhage. Pediat Res 16: 332 A

Baldes EJ, Farral WR, Haugen MC (1957) A forum on an ultrasonic method for measuring the velocity in blood. In: Kelly E (ed) Ultrasound in biology and medicine. Amer Inst Biol Sc, Washington, p 165

Bannister CM (1984) Ischemia and revascularization of the middle cerebral territory of the rat brain by manipulation of the blood vessels in the neck. Surg Neurol 21: 351–357

Bayliss WM (1902) On the local reactions of the arterial wall to changes of internal pressure. J Physiol (London) 28: 220–231

Beasley MG, Blau JN, Gosling RG (1979) Changes in internal carotid artery flow velocities with cerebral vasodilation and constriction. Stroke 10: 331–335

Berger EC, et al (1975) Intraoperative angiography in cerebral angiomas and aneurysms surgery. Acta Radiol Scr Diagn 16: 31–37 [Suppl 347]

Bernsmeier A, Siemons K (1952) Zur Messung der Hirndurchblutung bei intracraniellen Gefäßanomalien un deren Auswirkung auf den allgemeinen Kreislauf. Z. Kreisl-Forsch 41: 845

Betz E (1965) Zur Registrierung der lokalen Gehirndurchblutung mit Wärmeleitsonden. Pflügers Archiv 284

Bingham HG, Lichti E (1970) Use of ultrasound transducer (Doppler) to localize peripheral arteriovenous fistulae. Plast Reconstr Surg 46: 151–156

Blair WF, Greene ER, Omer GE (1981) A method for the calculation of blood in human digital arteries. J Hand Surg 6: 90–96

Bonnal J, Born JD, Hans P (1985) One-stage excision of high-flow arteriovenous malformations. J Neurosurg 62: 128–131

Boughner DR, Roach MR (1971) Effect of low frequence vibration on the arterial wall. Circ Res 29: 136–144

Brambilla G, Paoletti P, Radriguez y Baena R (1982) Extracranial-intracranial arterial bypass in the treatment of inoperable giant aneurysms of the internal carotid artery. Acta Neurochir (Wien) 60: 63–69

Brawley BW, Strandness Jr DE, Kelly WA (1967) The physiologic response to therapy in experimental cerebral ischemia. Arch Neurol 17: 180–187

Brescia MJ, Cimino JE, Appel K, Hurwich BJ (1966) Chronic hemodialysis using veinpuncture and surgically created arteriovenous fistula. N Engl J Med 275: 1083

Brihaye J, David M, Dilenge D (1963) Aspects physiopathologiques des fistules artérioveineuses cérébrales. Acta Chir Belg 62: 961–972

Büdingen JH, Gilsbach J, von Reutern GM (1978) Dopplersonographische Therapie- und Verlaufskontrolle einer katheteroccludierten Cavernosusfistel. Arch Psychiat Nervenkr 226: 19–27

— von Reutern GM, Freund HJ (1982) Dopplersonographie der extrakraniellen Hirnarterien. Thieme, Stuttgart New York

Brambilla G, Paoletti P, Rodriguez Y, Baena R (1982) Extracranial-intracranial arterial bypass in the treatment of inoperable giant aneurysma of the internal carotid artery. Acta Neurochir (Wien) 60: 63–69

Brisman R, Grossman BL, Correll JW (1970) Accuracy of transcutaneous Doppler ultrasonics in evaluating extracranial vascular disease. J Neurosurg 32: 529–533

Bruster DC, Schlaen HH, Raines JK, et al (1978) Rational management of the asymptomatic carotid bruit. Arch Surg 113: 927–930

Burki NK, Albert RK (1983) Noninvasive monitoring of arterial blood gases. Chest 83: 666–670

Cannon JA, Lobpreis EL, Herrold G, et al (1960) Experience with a new electromagnetic flowmeter for use in blood-flow determinations in surgery. Ann Surg 152: 635–646

Capon A, Martin PH, Noterman J (1969) Evaluation du débit sanguin local dans les tumeurs cérébrales à l'aide d'une méthode de thermodiffusion. Neurochirurgie 15: 569–574

Carter LP, White WL, Atkinson DR (1978) Regional cortical blood flow at craniotomy. Neurosurgery 2: 223–229

— Erspamer RBS, White WL, et al (1982) Cortical blood flow during craniotomy for aneurysm. Surg Neurol 17: 204–208

Cathignol D, Chapelon JY, Fourcade C (1978) Velocimètre Doppler à l'usage des petits vaisseaux. In: Le Microglo Biosigma Paris, 24–28 April 1978, pp 426–429

— Fourcade C (1980) Improvement of pulsed Doppler flowmeter for use in microvascular diagnosis. ICEU New Delhi A13: 18–20

— Chapelon JY, Mestas JL, Fourcade C (1983) Description et application d'un velocimètre ultrasonore Doppler pour les petits vaisseaux. Med & Biol Eng & Comput 21: 358–364

Chater N (1976) Surgical results and measurements of intraoperative flow in microsurgical anastomoses. In: Austin GM (ed) Microneurosurgical anastomoses for cerebral ischemia. Ch C Thomas, Springfield, Ill, pp 295–304

Colin CG, Chan YS (1977) Computed tomographic arteriography. J Comput Assist Tomogr 1: 165–168

Collice M, Fornari M, Porta M (1977) End-to-side anastomosis between carotid arteries and serial angiographic controls in rats.

In: Schmiedek P (ed) Microsurgery for stroke. Springer, New York Heidelberg Berlin, pp 159–162

— Scialfa G, Valsecchi F, et al (1980) Cortical artery pressure: preoperative and postoperative arteriographic finding in patients with internal carotid artery occlusion. In: Peerless SJ, McCormick CM (eds) Microsurgery for cerebral ischemia. Springer, New York Heidelberg Berlin, pp 138–141

Colon EJ, de Weerd JPC, Notermans SLH, Vingerhoets HM (1979) Reliability of Doppler sonography in extracranial cerebrovascular stenosis. Clin Neurol Neurosurg 81: 108–113

Cophignon J, Thurel C, Djindjian R, et al (1978) Cerebral arteriovenous malformations. Modern aspects of investigations and treatment. Prog Neurol Surg 9: 195–237

Cow D (1911) Some reactions of surviving arteries. J Physiol 42: 125–143

Crowell RM (1976) Electromagnetic flow studies of superficial temporal artery to middle cerebral branch artery bypass graft. In: Austin GM (ed) Microneurosurgical anastomoses for cerebral ischemia. Ch C Thomas, Springfield, Ill, pp 116–124

Cushing H, Bailey P (1928) Tumors arising from the blood vessels of the brain. Angiomatous malformations and hemangioblastomas. Ch C Thomas, Springfield, Ill

Dandy WE (1928) Arteriovenous aneurysm of the brain. Arch Surg 17: 190–243 (Chicago)

— (1928) Venous abnormalities and angiomas of the brain. Arch Surg 17: 715–793 (Chicago)

Day AL, Friedman WA, Sypert GW, et al (1982) Successful treatment of the normal perfusion pressure breakthrough syndrome. Neurosurgery 11: 625–630

Debrun G, Vinuela F, Fox A, et al (1982) Embolization of cerebral arteriovenous malformations with bucrylate. Experience in 46 cases. J Neurosurg 56: 615–627

Deruty R, Lapras C, Bret P, et al (1981) Embolisation peropératoire des malformations artérioveineuses cérébrales inextirpables. Tentative d'oblitération par un mélange à polymérisation retardée. A propos de deux cas. Neurochirurgie 27: 5–14

Dichgans J, Gottschalk M, Voigt K (1972) Arteriovenöse Dura-Angiome am Sinus transversus. Klinische Symptome, charakteristische arterielle Versorgung und häufige venöse Abflußstörungen. Zbl Neurochir 33: 1–18

Dobben GD, Valvassori GE, Mafee MF, Berninger WH (1979) Evaluation of brain circulation by rapid rotational computed tomography. Radiology 133: 105–111

Doppler C (1842) Über das farbige Licht der Doppelsterne und einiger anderer Gestirne des Himmels. Abh Kgl Böhm Ges d Wissensch Prag, pp 465–482

Drake CG (1979) Cerebral arteriovenous malformations: considerations for an experience with surgical treatment in 166 cases. Clin Neurosurg 26: 145–208

Du Boulay G, Edmonds-Seal J, Bostick T (1970) Effect of intermittent positive pressure ventilation upon cerebral angiography: quality of film and diameter of vessels including those in spasm. In: Taveras JM, Fischgold H, Dilenge D (eds) Recent advances in the study of cerebral circulation. Ch C Thomas, Springfield, Ill

Dunn V, Wing SD, Miller FJ, Koehler PR (1979) Hemodynamic studies using a CT scanner. CT 3: 173–179

Eden A (1984) Doppler techniques and neurosurgery. Neurosurg Rev 7: 193–197

Ekström-Jodal B, Häggendal E, Lindner LE, Nilsson NJ (1971/72) Cerebral blood flow autoregulation at high arterial pressures and different levels of carbon dioxide tension in dogs. Europ Neurol 6: 6–10

Elkin DC, Warren JV (1947) Arteriovenous fistulas. Their effect on the circulation. JAMA 134: 1524–1528

Enevoldsen E, Jensen FT (1978) Autoregulation and CO_2 responses of cerebral blood flow in patients with acute severe head injury. J Neurosurg 55: 857–864

Farrar JK, Gamache FW, Ferguson GG, et al (1981) Effects of profound hypotension on cerebral blood flow during surgery for intracranial aneurysms. J Neurosurg 55: 857–864

Fazekas JF, Bessmann AM, Cotsonas NJ, Alman RW (1953) Cerebral hemodynamics in cerebral arteriosclerosis. J Gerontol 8: 137

Fazio C (1970) The importance of the "intracerebral steal" in the pathogenesis of focal brain ischemia. In: Mayer JS, Reivich M, Lechner H, Eichhorn O (eds) Research on the cerebral circulation. Ch C Thomas, Springfield, Ill, pp 57–59

Fein JM, Mollinari G (1974) Experimental augmentation of regional cerebral blood flow by microvascular anastomoses. J Neurosurg 41: 421–426

Feindel W, Yamamoto YL, Hodge CP (1967) Intracarotid fluorescein angiography: a new method for examination of the epicerebral circulation in man. Canad med Ass J 96: 1

— — — (1971) Red cerebral veins and the cerebral steal syndrome. Evidence from fluorescein angiography and microregional blood flow by radioisotopes during excision of an angioma. J Neurosurg 35: 167–179

— — — (1971) The cerebral microcirculation in man: analysis by radioisotope microregional flow measurement and fluorescein angiography. In: Tindall GT, Wilkins RH, Keener EB (eds) Clinical neurosurgery, vol 18. Williams & Wilkins, Baltimore, pp 225–246

— (1975) The influence of cerebral steal: Demonstrations by fluorescin angiography and focal cerebral blood flow measurement. In: Pia HW, Gleave JRW (eds) Cerebral angiomas. Advances in diagnosis and therapy. Springer, Berlin Heidelberg New York

Ferguson GG (1972) Direct measurement of mean and pulsatile blood pressure at operation in human intracranial saccular aneurysms. J Neurosurg 36: 560–563

Fieschi O, Agnoli E, Galbo E (1963) Effects of carbon dioxide on cerebral hemodynamics in normal subjects and in cerebrovascular disease studies by carotid injection of radio-albumin. Circ Res 13: 436–447

Florey HW (1925) Microscopical observations on the circulation of the blood in the cerebral cortex. Brain 48: 43–64

Fog M (1934) Om piaarteriernes vasomotoriske reaktioner. Thesis, Copenhagen

Folkow B, et al (1971) The hemodynamic consequences of regional hypotension in spontaneously hypertensive and normotensive rats. Acta Physiol Scand 83: 532–541

Forbes HS (1928) Cerebral circulation: I. Observation and measurement of pial vessels. AMA Arch Neurol Psychiat 19: 751

Foreman JEK, Hutchinson KJ (1970) Arterial wall vibration distal to stenoses in isolated arteries of dog and man. Circ Res 26: 583–590

Freed D, Hartley CJ, Christman KD, Lyman RC, Agris J, Walker WF (1979) High-frequency pulsed Doppler ultrasound: a new tool for microvascular surgery. J Microsurg 1: 148–153

Freund JH (1965) Ultraschallregistrierung der Pulsation einzelner intrakranieller Arterien zur Diagnostik von Gefäßverschlüssen. Arch Psychiat Nervenkr 207: 247–253

Friedrich H, Hänsel-Friedrich G, Seeger W (1980) Intraoperative Doppler sonography of brain vessels. Neurochirurgia 23: 89–98 (Stuttgart)

Fuentes JM (1979) Cortical arterial pressure in extra-intracranial anastomosis. Acta Neurochir (Wien) [Suppl] 28: 272–274

Fujishima MK, Tanaka K, Fukiyama K, et al (1967) Brain hemodynamics in 11 cases with cerebral arteriovenous aneurysms. Brain Nerve (Tokyo, Jap) 19: 905–909 (Sept)

Gabrielsen TO, Greitz T (1970) Normal size of the internal carotid, middle cerebral and anterior cerebral arteries. Acta Radiol (Diagn) 10: 1–10

Gelber BR, Sundt TM (1980) Treatment of intracavernous and giant carotid aneurysms by combined internal carotid ligation and extra to intracranial bypass. J Neurosurg 52: 1–10

Gibbs FA (1933) Thermoelectric blood flow recorder in form of needle. Proc Soc Exp Biol Med 31: 141

— Gibbs EL, Lennox WG (1935) Changes in human cerebral blood flow consequent on alterations in blood gases. Am J Physiol 11: 557

Gilsbach JM (1983) Intraoperative Doppler sonography in neurosurgery. Springer, Wien New York

Gotoh F, Tazaki Y, Meyer JS (1961) Transport of gases through brain and their extravascular vasomotor action. Exp Neurol 4: 48

Grange RA, Hawkins TD, Samuel JR (1969) Influence of carbon dioxide tension on the angiographic appearance of intracranial tumors. Acta Radiol (Diagn) 9: 292–299

Green HD, Rapela CE, Conrad MC (1963) Resistance (conductance) and capacitance phenomena in terminal vascular bed. In: Hamilton WF, Dow P (eds) Handbook of physiology, Sec 2, Circulation, vol II, Amer Physiological Soc, Washington DC, pp 935–960

Green EC (1968) Anatomy of the rat. Hafner Publishing Co, New York London

Greene ER, Blair WF, Hartley CJ (1981) Noninvasive pulsed Doppler blood velocity measurements and calculated flow in human digital arteries. ISA Trans 20: 15–24

Greitz T (1956) A radiologic study of the brain circulation by rapid serial angiography on the carotid artery. Acta Radiol [Suppl] 140

Griffiths DPG, Cummins BH, Greenbaum R, et al (1974) Cerebral blood flow and metabolism during hypotension induced with sodium nitroprusside. Br J Anaesth 46: 671–679

Grubb B, Mills CD, Colacino JM, Schmidt-Nielsen K (1977) Effect of arterial carbon dioxide on cerebral blood flow in ducks. J Physiol 232: 596–608

Gündling P, Haneder J, Gaab MR (1985) Correlation between CBF and pCO, pO_2, pH, hemoglobin. blood pressure, age, and sex. In: Hartmann, Hoyer (eds) Cerebral blood flow and metabolism measurements. Springer, Berlin Heidelberg

Guidetti B, Delitala A (1980) Intracranial arteriovenous malformations. Conservative and surgical treatment. J Neurosurg 53: 149–152

Gygax P, Wiernsperger N (1983) Hypotension induced changes in cerebral microflow and EEG and their pharmacological alterations. Acta Med Scand [Suppl] 678: 29–35

Hacker H, Becker H (1977) Time controlled computed tomographic angiography. J Comput Assist Tomogr 1: 405–409

Häggendal E, Ingvar DH, Lassen NA, Milsson NJ, Norlén G, Wickbom I, Zwetnow N (1965) Pre- and postoperative measurements of regional cerebral blood flow in three cases of intracranial arteriovenous aneurysm. J Neurosurg 22: 1–6

Handa H, Moritake K, Nagata J, *et al* (1980) Intraoperative hemody-namic study by Doppler ultrasonic flowmeter in the extracranial-intracranial arterial bypass. In: Peerless SJ, McCormick CW (eds): Microsurgery for cerebral ischemia. Springer, New York Heidelberg Berlin, pp 99–105

Harders A, Gilsbach J (1985) Transcranial Doppler sonography and its application in extracranial-intracranial bypass surgery. Neurol Res 7: 129–141

Hardesty WH, Roberts B, Toole JF, *et al* (1960) Studies of carotid artery blood flow in man. N Engl J Med 263: 944–946

Harper AM (1965) The inter-relationship between P_3CO_2 and blood pressure in the regulation of blood flow through the cerebral cortex. Acta Neurol Scand 41: 94–103 [Suppl 14]

— Glass HI (1965) Effect of alterations in the arterial carbon dioxide tension on the blood flow through the cerebral cortex at normal and low arterial blood pressure. J Neurol Neurosurg Psychiatry 28: 449–452

Hartley CJ, Cole JS (1974) An ultrasonic pulsed Doppler system for measuring blood flow in small vessels. J Appl Physiol 37: 626–629

Hassler W, Gilsbach J, Gaitzsch J (1983) Results and value of immediate postoperative angiography after operation of ar-teriovenous malformation. Neurochirurgia 26: 146–148

Hassler W, Gilsbach J (1984) Intra- and perioperative aspects of the hemodynamics of supratentorial AV malformations. Acta Neu-rochir (Wien) 73: 35–44

Hauge A, Thoresen M, Walloe L (1980) Changes in cerebral blood flow during hyperventilation and CO_2-breathing measured trans-cutaneously in humans by a bidirectional, pulsed, ultrasound Doppler blood velocitymeter. Acta Physiol Scand 110: 167–173

Heilbrun MP, Olesen J, Lassen NA (1972) Regional cerebral blood flow studies in subarachnoid hemorrhage. J Neurosurg 37: 36–44

Heinz ER, Dubois P, Drayer B, Barrett W (1979) Dynamic computed tomography study of the brain. J Comput Assist Tomogr 3: 641–649

Hial SK (1974) Cerebral hemodynamics assessed by angiography. In: Newton TH, Poth DG (eds) Radiology of the skull and brain, Angiography vol II, Book 1. CV Mosby Co, St Louis, pp 1049–1085

Hitchon PW, Kassell NF, McDonnell DE (1979) The Doppler ultrasonic flowmeter as an adjunct to operative management of cerebral arteriovenous malformations. Surg Neurol 11: 345–347

Hoedt-Rasmussen K, Skinhoj E, Paulson O, Ewald J, Bjerrum JK, Fahrenkrug A, Lassen NA (1967) Regional cerebral blood flow in acute apoplexy. The "luxury perfusion syndrome" of brain tissue. Arch Neurol 17: 271–281

Höök O, Werkö L, Öhrberg G (1958) Intracranial arteriovenous aneurysms. A study of their effect on the cardiovascular system. Arch Neurol Psychiat 79: 622 (Chicago)

Huber P, Handa J (1967) Effect of contrast material, hypercapnia, hyperventilation, hypertonic glucose and papaverine on the diameter of the cerebral arteries—angiographic determination in man. Invest Radiol 2: 17–32

Hunter J (1935) Works. Palmer, London

Ingvar DH, Lassen NA (1962) Regional blood flow of the cerebral cortex determined by Krypton-85. Acta Physiol Scand 54: 325–338

Ito Z, Hen R, Nakajima K, *et al* (1980) Selection of completed stroke patients for STA-MCA anastomosis based on measurements of somatosensory evoked potential and CBF dynamics. In: Peerless SJ, McCormick CW, Microsurgery for cerebral ischemia. Sprin-ger, Berlin Heidelberg New York, pp 177–184

James IM, Millar RA, Purves MJ (1969) Observations on the extrinsic neural control of cerebral blood flow in the baboon. Circ Res 25: 77–93

Jorch G, Menge U (1985) Die Bedeutung des pCO_2 für die Hirndurchblutung in der Neonatologie. Eine Doppler-sonographische Untersuchung. Monatsschr Kinderheilkd 133: 38–42

Kalckreuth W von, Hillesheimer W, Reutern GM von (1985) Arteriovenous shunts and results of Doppler ultrasound in supraaortic vessels. In: Hartman, Hoyer (eds) Cerebral blood flow and metabolism measurement. Springer, Berlin Heidelberg, pp 562–565

Kalmus HP (1954) Electronic flowmeter system. Rev Sci Instr 25: 201

Keller H, Die cerebrovasculäre Doppler-Ultraschall-Untersuchung (cv-Doppler). Bull Schweiz Akad Med

Kety SS, Schmidt CF (1946) Effects of active and passive hyper-ventilation on cerebral blood flow, cerebral oxygen consumption, cardiac output, and blood pressure of normal young men. J Clin Invest 25: 107–119

— — (1948) The effects of altered arterial tensions of carbon dioxide and oxygen on cerebral blood flow and cerebral oxygen con-sumption in normal young men. J Clin Invest 27: 484–492

Kolin A (1936) An electromagnetic flowmeter. Principle of the method and its application to blood flow measurements. Proc Soc Exp Biol Med 35: 53

Kosugi Y, Goto T, Ikebe J, *et al* (1983) Sonic detection of intracranial aneurysm and AVM. Stroke 14: 37–42

Kreissmann A, Bollinger A (1978) Ultraschall-Doppler-Diagnostik in der Angiologie. G. Thieme, Stuttgart

Kristensen JK (1967) Ultrasonic pulse detection applied to the carotid and vertebral arteries. Scand J Thorac Cardiovasc Surg 1: 178–180

Kristiansen K, Krog J (1962) Electromagnetic studies on the blood flow through the carotid system in man. Neurology 12: 20–22

Krueger TP, Rockoff SD, Thomas LJ, Ommaya AK (1963) The effects of changes of end expiratory carbon dioxide tension on the normal cerebral angiogram. Am J Roentgenol Radium Ther Nucl Med 90: 506–511

Kurti XH, Zogu V, Petrela M, Leka LL (1984) Common carotid artery-internal vein aneurysms and fistula with cerebrovascular signs. A case report. Acta Neurochir (Wien) 71: 109–111

Lambertsen CJ, Semple SJG, Smyth MG, Gelfand R (1961) H^+ and pCO_2 as chemical factors in respiratory and cerebral circulatory control. J Appl Physiol 16: 473–484

Lassen NA (1964) Autoregulation of cerebral blood flow. Circ Res 15: 201–204 [Suppl 1]

— (1966) The luxury-perfusion syndrome and its possible relation to acute metabolic acidosis localized within the brain. Lancet 1: 1113–1115

— Agnoli A (1972) Upper limit of autoregulation of cerebral blood flow: on the pathogenesis of hypertensive enecephalopathy. Scand J Clin Lab Invest 30: 113

— Astrup J (1985) Cerebrovascular physiology. In: Fein JM, Flamm ES, Cerebrovascular surgery, vol I. Springer, New York Berlin Heidelberg Tokyo

— Ingvar DH (1961) The blood flow of the cerebral cortex determined by radio-active Krypton-85. Experientia 17: 42–43

Leksell L (1956) Echo-encephalography: detection of intracranial complications following head injury. Acta Chir Scand 110: 301–315

Lewis BM, *et al* (1960) Method for continuous measurement of cerebral blood flow in man by means of radioactive Krypton (Kr-79). J Clin Invest 39: 707

Little JR, Yamamoto YL, Feindel W, *et al* (1980) Cerebral blood flow in superficial temporal artery to middle cerebral anastomosis. In: Peerless SJ, McCormick CW (eds) Microsurgery for cerebral ischemia. Springer, New York Heidelberg Berlin, pp 59–60

Loeb C, Favale E (1962) Contralateral EEG abnormalities in intracranial arteriovenous aneurisms. Arch Neurol 7: 121

Loop JW, Foltz EL (1966) Applications of angiography during intercranial operation. Acta Radiol (Diagn) 5: 363–367

Lovick AHJ, Pickard JD, Goddard BY (1982) Prediction of late ischemic complications after cerebral aneurysm surgery. Use of a mobile microcomputer system for the measurement of pre-, intra-, and postoperative cerebral blood flow. Acta Neurochir (Wien) 63: 37–42

Luessenhop AJ, Gennarelli TA (1977) Anatomical grading of supratentorial arterio-venous malformations for determining operability. Neurosurgery 1: 30–35

— Ferraz FM, Rosa L (1982) Estimate of the incidence and importance of circulatory breakthrough in the surgery of cerebral arteriovenous malformations. Neurol Res 4: 177–190

Markwalder TM, Grolimund P, Seiler RW, Roth F, Aaslid R (1984) Dependency of blood flow velocity in the middle cerebral artery on end-tidal carbon dioxide partial pressure—a transcranial ultrasound Doppler study. J Cerebral Blood Flow and Metabolism 4: 368–372

Matjasko MJ, Williams JP, Fontanilla M (1975) Intraoperative use of Doppler to detect successful obliteration of carotid-cavernosus fistulas. J Neurosurg 43: 634–636

McCormick WF (1966) Report on the cooperative study on intracranial aneurisms: the pathology of vascular ("arteriovenous") malformations. Neurosurg 24: 807–816

McDonald DA (1974) Blood flow in arteries. Williams & Wilkins Co, Baltimore, p 3

Melamed E, Lavy S, Bentin S, Cooper G, Rinot Y (1980) Reduction in regional cerebral blood flow during normal aging in man. Stroke 1: 31–34

Menon D, Weir B (1979) Evaluation of cerebral blood flow in arteriovenous malformations by the xenon 133 inhalation method. Can J Neurol Sci 6: 411–416

Merory J, du Boulay GH, Marshall J (1979) Cerebral blood flow following aneurysmal surgery after subarachnoid hemorrhage. Acta Neurochir (Wien) 46: 180

Meyer JS, Fukuuchi Y, Kanda T, Shimazu K, Hashi K (1972) Regional cerebral blood flow measured by intracarotid injection of hydrogen. Comparison of regional vasomotor capacitance from cerebral infarction versus compression. Neurology 22: 571–584

— — Shimazu K, Ohuchi T, Ericsson AD (1972) Abnormal hemispheric blood flow and metabolism in cerebrovascular disease. II. Therapeutic trials with 5% CO_2 inhalation, hyperventilation and intravenous infusion of Tham and mannitol. Stroke 3: 157–167

Meyer JS, Ishihara N, Deshmukh VD, Naritomi H, Sakai F, Hsu M-Cö, Pollack P (1978) Improved method for noninvasive measurements of regional cerebral blood flow by Xenon inhalation, Part I and II. Stroke 3: 195–210

Miller JD, Jawald K, Jennett B (1977) Safety of carotid ligation and its role in the management of intracranial aneurisms. J Neurol Neurosurg Psychiatry 40: 64–72

Mills CJ (1972) Measurement of pulsatile flow and flow velocity. In: Bergel DH (ed) Cardiovascular fluid dynamics, vol 1. Academic Press, London, pp 51–90

Miyazaki M, Kato K (1965) Measurement of cerebral blood flow by ultrasonic Doppler technique. Jpn Circ J 29: 375–382

Mol JMF, Rijcken WJ (1974) In: Reneman RS (ed) Cardiovascular applications of ultrasound. Elsevier North-Holland-Excerpta Medica, Amsterdam, pp 305–314

Moniz E (1934) L'angiographie cérébrale. Masson et Cie, Paris

Moritake K, Handa H, Yonekawa Y, Nagata I (1980) Ultrasonic Doppler assessment of hemodynamics in superficial temporal artery—middle cerebral artery anastomosis. Surg Neurol 13: 249–257

Mullan S, Brown FD, Patronas NJ (1979) Hyperemic and ischemic problems of surgical treatment of arterio-venous malformations. J Neurosurg 51: 757–764

Mullaart RA, Daniels O, Hopmann JCW, *et al* (1982) Ultrasound detection of congenital arteriovenous aneurysm of the great cerebral vein of Galen. Eur J Pediatr 139: 195–198

Murphy JP, Cerebrovascular disease. Year Book Publishers, Chicago 1954

Newton TH, Potts DG (1974) Radiology of the skull and brain; Angiography Vol II, book 4, part IV. Comparative vascular anatomy. The CV Mosby Co, St Louis, pp 2763–2786

Nickel R (1984) Lehrbuch der Anatomie der Haustiere, Band III: Kreislaufsystem, Haut und Hautorgane. Verlag Paul Parey, Berlin Hamburg

Nies JMM (1976) The haemodynamic effect of an intracranial arteriovenous anomaly. A doppler-haematotachographic study. Clin Neurol Neurosurg 79: 29–45

Nilsson GE, Tenland T, Öberg PA (1980) A new instrument for continuous measurement of tissue blood flow by light beating spectroscopy. IEEE Trans Biomed Eng BME 27: 12–29

Norlen G (1949) Arteriovenous aneurysms of the brain. Report of ten cases of total removal of the lesion. J Neurosurg 6: 475–494

Norman D, Axel L, Berninger WH, Edwards MS, Conn CE, Redington RW, Cox L (1981) Dynamic computed tomography of the brain: Techniques, data analysis, and applications. AJR 136: 759–770

Nornes H (1968) Longtime implanted electromagnetic flow probes in man. Observations during graded occlusion for internal carotid artery aneurisms. In: Cappelen C, New findings in blood flowmetry, Universiteitsforlaget, Oslo, pp 215–219

— (1972) Hemodynamic aspects in the management of carotid-cavernosus fistula. J Neurosurg 37: 687–694

— (1976) Electromagnetic blood flowmetry in small vessel surgery: an experimental study. Scand J Thor Cardiovasc Surg 10: 144–148

— Wickeby P (1977) Cerebral arterial blood flow and aneurysm surgery, Part 1: Local arterial flow dynamics. J Neurosurg 47: 810–818

— Grip A, Wikeby P (1979) Intraoperative evaluation of cerebral hemodynamics using directional Doppler technique, Part 1: Arteriovenous malformations. J Neurosurg 50: 145–151

— — — (1979) Intraoperative evaluation of cerebral hemodynamics using directional Doppler technique, Part 2: Saccular aneurisms. J Neurosurg 50: 570–577

— — (1980) Hemodynamic aspects of cerebral arteriovenous malformations. J Neurosurg 53: 456–464

Obrist WD, Thomson HK, Wang HS, Wilkinson WE (1975) Regional cerebral blood flow estimate by 133-Xenon inhalation. Stroke 6: 245–256

Oeconomos D, Kosmaoglou D, Prossalentis A (1969) rCBF studies in patients with arteriovenous malformations of the brain. In: Cerebral blood flow clinical and experimental results. Springer, Berlin, pp 146–148

Okabe T, Meyer JS, Okayasu H, Harper R, Rose J, Grossman RG, Centeno R, Tachibana H, Lee YY (1983) Xenon-enchanced CT CBF measurements in cerebral AVMs before and after excision. Contribution to pathogenesis and treatment. J Neurosurg 59: 21–31

Olinger CP, Wasserman JF (1977) Electronic stethoscope for detection of cerebral aneurysm, vasospasm and arterial disease. Surg Neurol 8: 298–312

Olivecrona H, Rives J (1948) Arteriovenous aneurysms of the brain: their diagnosis and treatment. Arch Neurol Psychiat 59: 567–602

Palvölgyi R (1969) Regional cerebral blood flow in patients with intracranial tumors. J Neurosurg 31: 149

Parkinson D (1958) Cerebral arteriovenous aneurysms: surgical management. Canad J Surg 1: 313–324

— (1969) Rapid serial simultaneous biplane stereoscopic angiography; an aid in the surgical management of cerebral arteriovenous malformations. Clin Neurosurg 16: 179–184

— (1973) Carotid cavernous fistula: direct repair with preservation of the carotid artery. Technical note. J Neurosurg 38: 99–106

Patterson JL Jr, Heyman A, Battey LL, Ferguson RW (1955) Threshold of response of the cerebral vessels of man to increase in blood carbon dioxide. J Clin Invest 34: 1857–1864

Paulson OB, Lassen NA, Skinhoj E (1970) Regional cerebral blood flow in apoplexy without arterial occlusion. Neurology 20: 125–138 (Minneap)

Peerless SJ (1982) Comments on Day Al, Friedman WA, Sypert GW, et al, Successful treatment of the normal perfusion pressure breakthrough syndrome. Neurosurgery 11: 629–630

Pertuiset B, Ancri D, Lienhart A (1981) Profound arterial hypotension (MAP 50 mmHg) induced with neuroleptanalgesia and sodium nitroprusside (series of 531 cases). Reference to vascular autoregulation mechanism and surgery of vascular malformations of the brain. In: Krayenbühl H et al (eds) Advances and technical standards in neurosurgery, vol 8. Springer, Wien New York, pp 75–122

— — Sichez JP, et al (1983) Radical surgery in cerebral AVM tactical procedures based upon hemodynamic factors. In: Krayenbühl H et al (eds) Advances and technical standards in neurosurgery, vol 10. Springer, Wien New York, pp 81–143

Phillips DJ, Baker DW, Strandness DE Jr (1982) Combined echo-Doppler (Duplex) imaging. In: Bernstein EF (ed) Noninvasive diagnostic techniques in vascular disease, 2nd edn. CV Mosby, St Louis, pp 272–280

Pia HW, Gleave JRW, Grote E, et al (eds) (1975) Cerebral angiomas: Advances in diagnosis and therapy. Springer, New York

Pourcelot L (1974) Applications cliniques de l'examen Doppler transcutane. Les Colloques de l'Institut National de la Santé et de la Recherche Médicale Inserm 34: 213–240

Power D'Arcy (1888) Angioma of the cerebral membranes. Tr Path Soc Lond 39-4: 77–79

Prosenz P, Heiss WD, Kvicala V, Tschabitscher H (1971) Contribution to the hemodynamics of arterial venous malformations. Stroke 2: 279

Raichle ME, Stone H (1972) Cerebral blood flow autoregulation and graded hypercapnia. Europ Neurol 6: 1–5

Rapela CE, Green HD (1964) Autoregulation of canine cerebral blood flow. Circ Res 15: 205 [Suppl 1]

— — Denison AB Jr (1967) Baroreceptor reflexes and autoregulation of cerebral blood flow in the dog. Circ Res 21: 559

Reivich M (1964) Arterial pCO_2 and cerebral hemodynamics. Am J Physiol 206: 25–35

— (1972) Regional cerebral blood flow in physiologic and pathophysiologic states. Prog Brain Res 35: 191–228

Reneman RS, Hoeks PG (1982) Doppler ultrasound in the diagnosis of cerebrovascular disease. J Wiley & Sons, Chichester New York Brisbane Toronto Singapore

Reutern GM von, Büdingen HF, Nenerici M, Freund HJ (1976) Diagnose und Differenzierung von Stenosen und Verschlüssen der Arteria carotis mit der Doppler-Sonographie. Arch Psychiat Nervenkr 222: 191–207

Ring BA (1967 b) Intraluminal diameters of the intracranial arteries. Vasc Surg 1: 137–151

Riva C, Ross B, Benedek GB (1972) Laser doppler measurements of blood flow in papillary tubes and retinal arteries. Intest Ophthalmol 11: 936–944

Rockoff SD, Doppman J, Krueger TP, Thomas TJ, Ommaya AK (1966) Altered opacification of the external carotid circulation by changes of end expiratory carbon dioxide tension. Invest Radiol 1: 123–128

Roy CS, Sherrington CS (1890) On the regulation of the blood-supply of the brain. J Physiol 11: 85 (London)

Rüftenacht D (1986) personal communication

Sagawa K, Guyton AC (1961) Pressure-flow relationships in isolated canine cerebral circulation. Am J Physiol 200: 711

Samson DS, Boone S (1978) Extracranial-intracranial (EC-IC) arterial bypass: past performance and current concepts. Neurosurgery 3: 79–86

Satomura S (1959) Study of the flow patterns in peripheral arteries by ultrasonics. J Acoust Soc Jpn 15: 151–158

Schieve JF, Wilson WP (1953) The influence of age, anesthesia and cerebral arteriosclerosis on cerebral vascular activity to CO_2. Am J Med 15: 171–174

Schmidt CF, Hendrix JP (1937) The action of chemical substances on cerebral blood vessels. Res Publ Assoc Res Nervous Mental Disease 18: 229–276

Schmidt RG, Thews G (eds) (1985) Physiologie des Menschen. 22nd corrected edn. Springer, Berlin

Scott BB, McGillicuddy JE, Seeger JF, Kindt GW, Giannotta SL (1978) Vascular dynamics of an experimental cerebral arteriovenous shunt in the primate. Surg Neurol 10: 34–38

Seeger W (1980) Microsurgery of the brain. Anatomical and technical principles, vol 1 and 2. Springer, Wien New York

— (1985) Differential approaches in microneurosurgery of the brain. Springer, Wien New York

Sekhar LN, Wasserman JF (1984) Noninvasive detection of intracranial vascular lesions using an electronic stethoscope. J Neurosurg 60: 553–559

Shapiro W, Wassermann AJ, Patterson JL (1965) Human cerebrovascular response time to evaluation of arterial carbon dioxide tension. Arch Neurol 13: 130–138

— — — (1966) Mechanisms and pattern of human cerebrovascular regulation after rapid changes in blood CO_2 tension. J Clin Invest 45: 913–922

Shenkin HA, Spitz EB, Grant FC, Kety SS (1948) Physiologic studies of arteriovenous anomalies of the brain. J Neurosurg 5: 165–172

Shima J, Nichida M, Okade Y (1983) Cerebral arterial blood flow measured with an electromagnetic flow meter during surgery. Neurol Med Chir 23: 343–348 (Tokyo)

Siesjö BK, Nilsson L (1971) Energy metabolism of the brain at reduced cerebral perfusion pressure in arterial hypoxaemia. In: Brierley JB, Meldrum BS (eds) Brain hypoxia. Heinemann, London

— Berntman L, Nilsson B (1980) Regulation of microcirculation in the brain. Microvasc Res 19: 158–170

Simkins TE, Stehbens WE (1974) Vibrations recorded from the adventitial surface of experimental aneurysms and arteriovenous fistulas. Vascular Sur 8: 153–165

Sivakoff M, Nouri S (1982) Diagnosis of vein of Galen arteriovenous malformation by two-dimensional ultrasound and pulsed Doppler method. Pediatrics 69: 84–86

Smith AL, Neufeld GR, Ominsky AJ, Wollman H (1971) Effect of arterial CO_2 tension in cerebral blood flow, mean transit time, and vascular volume. J Appl Physiol 31: 701–707

Sokoloff L (1960) The effect of carbon dioxide on the cerebral circulation. Anesthesiology 21: 664–673

Spencer MP, Reid JM (eds) (1981) Cerebrovascular evaluation with Doppler ultrasound. M Nijhoff, The Hague Boston London

Spetzler RF, Wilson CB, Weinstein P (1978) Normal perfusion pressure breakthrough theory. Clin Neurosurg 25: 651–672

— Schuster H, Roski RA (1980) Elective extracranial-intracranialarterial bypass in the treatment of inoperable giant aneurysms of the internal carotid artery. J Neurosurg 53: 22–27

Stattin S (1973) Changes in meningeal and cerebral circulation in subarachnoid haemorrhage: a preliminary report. Neuroradiology 5: 20–23

Staubesand J (1978) Matrix Vesikel and Mediodysplasie: ein neues Konzept zur formalen Pathogenese der Varikose. Phlebol Proktol 7: 109–140

— Fischer N (1979) Collagen dysplasie and matrix vesicles: researches with the electron microscope into the problem of socalled "weakness of the vessel wall." Path Res Pract 165: 374–391

— (1980) Hämodynamische Fehlbelastung — Risikofaktor für die Gefäßwand. Ergebn Angiol 20: 7–23

— Fischer N (1980) The ultrastructural characteristics of abnormal collagen fibrils in various organs. Connect Tissue Res 7: 21–217

Stephens HW Jr (1978) Electromagnetic blood flowmetry in microvascular anastomosis. In: Fein JM, Reichman OH (eds) Microvascular anastomoses for cerebral ischemia. Springer, New York Heidelberg Berlin, pp 181–194

Stephens HW (1980) Measurement of intracranial arterial pressure in patients undergoing extracranial to inctracranial microsurgical anastomoses for cerebrovascular ischemia. In: Peerless SJ, McCormick CW (eds) Microsurgery for cerebral ischemia. Springer, New York Heidelberg Berlin, pp 252–256

Stephenson HE, Lichti E (1971) Application of the Doppler ultrasonic flowmeter in the surgical treatment of arteriovenous fistula. Am Surg 37: 537–538

Strandgaard S, et al (1973) Autoregulation of brain circulation in severe arterial hypertension. Br Med J 1: 507–510

Straßburg HM, Niederhoff H, Sauer M (1982) Die dopplersonographische Registrierung intrakranieller Gefäße beim Säugling. Monatsschr Kinderheilk 130: 608–612

Symon L (1968) Experimental evidence for intracerebral steal following CO_2 inhalation. Scand J Clin Lab Invest [Suppl] 102, XIII A

Tanaka T, Riva C, Ben-Sira J (1974) Blood velocity measurements in human retinal vessels. Science 186: 830–831

Tanaka K, Tadaatsu N, Yoneda S, et al (1981) Ultrasonic evaluation of superficial temporal artery-middle cerebral artery anastomosis. Stroke 12: 803–807

Thoresen M, Walloe L (1979) Changes in cerebral blood flow during hyperventilation and CO_2-breathing measured by ultrasound. J Physiol 298: 53–54

Tindall Gut, Odom GL, Dillon ML, et al (1963) Direction of blood flow in the internal and external carotid arteries following occlusion of the ipsilateral common carotid artery: observation in 19 patients. J Neurosurg 20: 985–994

Tominaga S, Strandgaard S, Uemura K, Ito K, Kutsuzawa, Lassen NA, Nakamura T (1976) Cerebrovascular CO_2 reactivity in normotensive and hypertensive man. Stroke 7: 507–510

Traupe H, Heiss WD, Hoeffken W, et al (1980) Perfusion patterns in CT transit studies. Neuroradiology 19: 181–191

Tsuda Y, Kimura K, Yoneda S, Etani H, Nakamura M, Matsumoto M, Abe H (1983) Bi-hemispheric CBF and its CO_2 reactivity of TIAs and completed strokes in ICA occlusions. Neurol Res 5: 1–15

— — — — Asai T, Nakamura M, Abe H (1983) Cerebral blood flow and CO_2 reactivity in transient ischemic attacks: comparison between TIAs due to the ICA occlusion and ICA mild stenosis. Neurol Res 5: 17–37

Tulleken CAF, Abraham J (1975) The influence of changes in arterial CO_2 and blood pressure on the collateral circulation and the regional perfusion pressure in monkeys with occlusion of the middle cerebral artery. Acta Neurochir (Wien) 32: 161–173

Virchow R (1863) Die krankhaften Geschwülste. Vorlesung, Vol 3, 25. Berlin, pp 345–406

Wallace JM, Nashold BS Jr, Slewka AP (1965) Hemodynamic effects of cerebral arteriovenous aneurysms. Circulation 31: 696

Walter W (1975) The influence of the type and localization of angioma on the clinical syndrome. In: Pia HW, Gleave JRW, Grote E, Zierski J (eds) Cerebral angiomas. Springer, Berlin Heidelberg New York, pp 27–33

Waltz AG (1970) Effect of $PaCO_2$ on blood flow and microvasculature of ischemic and nonischemic cerebral cortex. Stroke 1: 27–37

Wasenko JJ, Cacayorin ED, Petro GR, Salibi NA, Hodge CH, Modesti LM, Kind RB (1985) Dynamic computed tomography: intracranial applications. Neurosurgery 16: 573–578

Wassermann AJ, Patterson JC (1961) The cerebral vascular response to reduction in arterial carbon dioxide tension. J Clin Invest 40: 1297–1303

Watts SH (1980) The suture of blood vessels. Implantation and transplantation of vessels and organs. An historical and experimental study. Johns Hopkins Hosp Bull 18: 153–179

Weinstein PR, Mehdorn HM, Spetzler RF, Telles DA (1980) Arterial dilatation and augmentation of blood flow in experimental arteriovenous fistulas. In: Peerless SJ, McCormick CW (eds) Microsurgery for cerebral ischemia. Springer, New York Heidelberg Berlin, pp 213–226

Weir B (1981) Value of immediate postoperative angiography following aneurysm surgery. J Neurosurg 54: 396–398

Wetterer E (1937) Eine neue Methode zur Registrierung der Blutströmungsgeschwindigkeit am uneröffneten Gefäß. Biol 98: 26

Whalen WJ, Ganfield R, Nair P (1970) Effects of breathing O_2 or $O_2 + CO_2$ and of the injection of neurohumors on the PO_2 of cat cerebral cortex. Stroke 1

White RJ, Albin MS, Yashon D, Verdura J, Austin JC, Austin PE, Demian YK (1976) Autoregulation in the isolated brain during profound hyperthermia and hypercarbia. In: Ross Russell RW (ed) Brain and blood flow. Pitman, London, pp 209–214

Whiteshell R, Asiddao C, Gollman D, Jablonski J (1981) Relationship between arterial and peak expired carbon dioxide pressure during anesthesia and factors influencing the difference. Anesth Analg 60: 508

Wiesländer JB (1980) Blood flow in small arteries after end-to-end and end-in-end anastomoses: an experimental quantitative comparison. J Microsurg 2: 121–125

Wilkins DG, Cummins BH, Griffith HB, et al (1972) Repeated measurements of cerebral blood flow during intracranial surgery. Lancet 2: 402–403

Williams PC, Stern MD, Bowen PD, Brooks Ra, Hammock MK, Bowman RL, Di Chiro G (1980) Mapping of cerebral cortical strokes in rhesus monkeys by laser Doppler spectroscopy. Med Res Eng 13: 3–5

Wilson CB, Domingue J (1979) Microsurgical treatment of intracranial vascular malformations. J Neurosurg 51: 446–454

— Stein BM (eds) (1984) Intracranial arteriovenous malformations. Williams & Wilkins, Baltimore London

Wing SD, Anderson RE, Osborn AG (1980) Dynamic cranial computed tomography: preliminary results. AJR 134: 941–945

Wintermantel E (1982) The thermic vascular anastomosis (TVA), Part 2: Microvascular auscultation applied to thermic vascular anastomoses. Acta Neurochir (Wien) 65: 277–287

Wolff HG, Lennox WG (1930) Cerebral circulation: XII. Effect on pial vessels of variations in the oxygen and carbon dioxide content of the blood. Arch Neurol Psychiat 23: 1097

Wollschlaeger B, Wollschlaeger G (1966) Anterior cerebral/internal carotid artery and middle cerebral/internal carotid artery ratios. Acta Radiol Diagnosis 5: 615–620

Worthington G, Schenk JR, et al (1957) The regional hemodynamics of experimental acute arteriovenous fistulas. Surg Gynecol Obstet December: 733–740

— — et al (1960) The regional hemodynamics of chronic experimental acute arteriovenous fistulas. Surg Gynecol Obstet January: 44–50

Wright RL (1968) Intraaneurysmal pressure reduction with carotid occlusion: observations in three cases of middle cerebral aneurysms. J Neurosurg 29: 139–142

Wüllenweber R (1965) Observation concerning autoregulation of central blood flow in man. Acta Neurol Scand 41: 111–115 [Suppl 14]

— (1968) "Intracerebral steal" in man recorded by a heat clearance technique. Scand J Lab & Clin Invest [Suppl 14]

Yamada S (1982) Arteriovenous malformations in the functional area: surgical treatment and regional cerebral blood flow. Neurol Res 4: 283–322

Yamamoto YL, Philips KM, Hodge CP, et al (1971) Microregional blood flow changes in experimental cerebral ischemia. Effects of arterial carbon dioxide studied by fluorescein angiography and xenon-133 clearance. J Neurosurg 35: 155–166

Yaşargil MG, Jain KK, Antic J, et al (1976) Arteriovenous malformation of the anterior and middle portion of the corpus callosum: microsurgical treatment. Surg Neurol 5: 67

— — – et al (1976) Arteriovenous malformations of the splenium of the corpus callosum: microsurgical treatment. Surg Neurol 5: 5

— Yonekawa Y (1978) Experience with the STA-MCA anastomosis in 46 cases. In: Fein JM, Reichman OH (eds) Microvascular anastomoses for cerebral ischemia. Springer, New York Heidelberg Berlin, pp 272–277

Yonekawa Y, Yaşargil MG (1976) Extra-intracranial arterialanastomosis: clinical and technical aspects. In: Krayenbühl H et al (eds), Advances and technical standards in neurosurgery, vol 3. Springer, Wien New York, pp 47–78

Zierler KL (1962) Theoretical basis of indicator-dilution methods for measuring flow and volume. Circ Res 10: 396–407

Subject Index

Aliasing 40, 41
Angiography, correlations with TCD 49, 52, 53, 61–64, 66, 67, 94, 95
—, CO$_2$-activity 79, 80, 94, 95
—, findings in angiomas 33–37, 94
—, postoperative 33–37
Angioma, Galenic 58, 59
—, angioarchitecture of 4
—, angiographical findings 33–37
—, hemodynamics in 5, 6
Angle of insonation 41
Animal model of Scott 7
— — — Spetzler 8
— models of AVM 7–9
Arteriovenous fistula, experimental, application of 7–32
—, bilateral 9
—, direct (T-type) 8, 10–13
—, indirect (H-type) 8, 13–32
Arteries, cerebral, diameter of 49, 51
—, —, TCD identification of 41
Autoregulation in angiomas 5
— — experimental AV-fistulas 26–28, 32, 48, 50
—, mechanisms of cerebrovascular 3, 44, 101, 103

Bayliss effect 103
Blood brain barrier 28, 29
— distribution, alterations of 79, 80
— flow rate—see flow rate
— —, regional cerebral—see CBF
— — velocity—see flow velocity
— viscosity 2
Brain death, TCD in 102, 105
Breakthrough phenomenon 6
Bruits in AVM 48, 49
— — experimental AV-fistulas 22, 24

Cerebral blood flow (CBF), regional—see CBF
CBF in angiomas 5, 52, 54
— — brain infarction 103, 107
— — head trauma 107
—, normal values 52, 54, 61, 103, 104
Cerebrovascular resistance 9, 69, 80—see resistance index
Circulatory breakthrough 2, 125, 126
CO$_2$-reactivity, angiographical 93–95
—, capacity of (TCD) 83

— in angioma feeders (TCD) 86–93, 107
—, measurement of (TCD) 79–82
—, normal values (TCD) 82–86
—, terminology 83
pCO$_2$, measurement of 80
Compression tests 41, 55–58

Doppler equation 38, 41
— effect 38, 41
— frequency 38, 41
— principle 38–41
— shift 38, 40, 47
— sonography, extracranial in AVM 38–60
— —, intraoperative 109–125
— —, transcranial—see TCD

Electromagnetic measurements 9, 10–32, 110
Electronic auscultation 39, 109
Embolization of AVM 66, 67, 96–100
Evans blue method 32

Fahraeus-Lindquist effect 2
Fistula, arteriovenous—see arteriovenous fistula
— occlusion test 12, 13, 15, 16
Flow rate, calculation of 52–54, 61
— — in angioma feeders 53, 54, 61
— — — AV-fistulas 10, 15, 17, 19, 20
— —, measurement of—see electromagnetic measurements
— velocity in angioma feeders 45–49, 77, 78
— — — basal cerebral arteries 42, 43, 77
— — — experimental AV-fistulas 10, 12, 14, 15, 19
Fluorescein angiography 109

Galenic angioma 58, 59

Hagen-Poiseuille's law 3
Heat emission method 39, 103, 109
Hemodynamics in normal vessels 2–4
— — angioma feeders 5, 6, 77–79
— — the angioma 5, 6, 77–79
Hemorrhage from AVM 51, 74–76, 96, 100, 101
History of AVM surgery 1–2
Hypercapnia, induced 79–85
Hyperperfusion 2, 126, 127

Hypertension, arterial, induced 22, 25–29, 43, 44, 50
—, intracranial 59
Hypertensive encephalopathy 29, 32
Hypocapnia, induced 79–85
Hypotension, arterial, induced 22, 25–29, 50, 58

Identification of basal cerebral arteries (TCD) 41
Index of resistance 9, 31, 42, 80, 85
Infrared analysis of pCO$_2$ 80, 81
Insonation angle 41
Interhemispheric difference of flow velocity 55, 88, 107
— steal in angiomas 55, 66, 88, 107
Intracerebral steal in angiomas 55, 66, 88, 107
Intracranial hypertension, pulse waveform in 77, 59
Intravascular pressure measurements 114–116, 118, 119

Kety-Schmidt (NO$_2$) method 39, 103, 107

Laminar flow 2, 3
Laser Doppler method 10, 24, 26–28, 30–32

Microcirculation, cortical 10, 24, 26–28, 30–32
Moya-moya syndrome 60

Normal perfusion pressure breakthrough theory 2, 5, 6, 126, 127

Pourcelot's formula 9
Pressure in angioma feeders 114–116, 118, 119
Pulsatility in angioma feeders 75, 76, 78
— — — veins 60
Pulse echo method 39
— repetition frequency 39, 40

Recording depth 39–41
Resistance index 9, 31, 42, 80, 85
—, peripheral vascular—see resistance index
Reynold's value 2, 3

Sample volume 40
Scott's animal model 7–9

Spetzler's animal model 7–9
Stagnating artery 33–37
Steal phenomenon, intracerebral 88, 90,
    107
— —, interhemispheric 33–37, 55, 57
Stenosis in experimental AV-fistula 11–13,
    17–19
Subarachnoid hemorrhage from AVM 51,
    59, 101

TCD and CO$_2$-reactivity—see CO$_2$-reactivity

—, examination technique 41
—, identification of arteries 39–42
— in angioma patients 45–108
— methodology 39–42
—, normal values 42
Thrombosis in angioma veins 126, 127
Transcranial Doppler Sonography—see
    TCD
Turbulency 22, 24, 48, 49

Ultrasonic beam 40, 41

Vasospasm in angioma feeder 49, 51, 59, 101
Vein, angioma draining 60
Velocity ratio, diastolic—systolic 78
Vessel occluder 20, 21
Vibrations of vessel walls 22, 24
Viscosity—see blood viscosity

Willis, circle of 41, 55–58

Neurosurgical Applications of Transcranial Doppler Sonography

By **A. Harders,** M. D.,
Neurochirurgische Universitätsklinik,
Freiburg i. Br., F. R. G.

1986. 109 figures. X, 134 pages.
Soft cover DM 58,–, öS 406,–
ISBN 3-211-81938-X

In 1981 Dr. Rune Aaslid developed a transcranial Doppler device with a pulsed sound emission of 2 MHz, which enabled blood flow velocities to be measured in the large branches of the circle of Willis. With this innovation, it has become possible to record atraumatically and repeatedly the intracranial hemodynamic changes in neurovascular diseases.

The book describes the hemodynamic principles in cerebral vascular circulation and the factors which can effect the blood flow velocities (such as collateral circulation, diameter of the vessels, vascular resistance, arterial partial CO_2 pressure, autoregulatory factors, and position of the body). Normal values of blood flow velocities and the changes under physiological deviations are measured by transcranial Doppler technique. For patients suffering from subarachnoid hemorrhage, individual time courses of velocity changes are evaluated and the application in clinical routines is stressed: Better defined timing of angiography, surgery and postoperative hypertension therapy has significantly reduced the incidence of delayed ischemic deficits.

Patients indicating for extracranial-intracranial bypass surgery, as well as the postoperative changed hemodynamics are also investigated. The contribution of the bypass to the brain circulation can be tested by compression tests. The "activity" of an angioma and the influence of superselective embolization procedures for arteriovenous malformations are described.

Furthermore, cerebro-vascular blood flow arrest in brain death patients, can clearly be seen without angiography by evaluating a reverberating flow pattern. The book gives an account of the role of a still very young but exciting technique in diagnostic and therapeutic procedures of cerebral vascular disease based upon three years of experience at the Neurosurgical Department of the University of Freiburg.

Transcranial Doppler Sonography

Edited by **R. Aaslid,** Ph. D.,
Director, Cardiovascular Research,
Institute of Applied Physiology and Medicine,
Seattle, Washington, U.S.A.

1986. 94 figures. XI, 177 pages.
Soft cover DM 68,–, öS 476,–
ISBN 3-211-81935-5

Contents: A. Eden: The Beginnings of Doppler. – P. Grolimund: Transmission of Ultrasound Through the Temporal Bone. – R. Aaslid: The Doppler Principle Applied to Measurement of Blood Flow Velocity in Cerebral Arteries. – R. Aaslid: Transcranial Doppler Examination Techniques. – R. Aaslid, K.-F. Lindegaard: Cerebral Hemodynamics. – K.-F. Lindegaard, R. Aaslid, H. Nornes: Cerebral Arteriovenous Malformations. – J. M. Gilsbach, A. Harders: Comparison of Intraoperative and Transcranial Doppler. – R. W. Seiler, R. Aaslid: Transcranial Doppler for Evaluation of Cerebral Vasospasm. – A. Harders: Monitoring Hemodynamic Changes Related to Vasospasm in the Circle of Willis After Aneurysm Surgery. – E. B. Ringelstein: Transcranial Doppler Monitoring. – T. Lundar: Transcranial Doppler in the Study of Cerebral Perfusion During Cardiopulmonary Bypass. – Subject Index.

From the Foreword by M. P. Spencer, M. D., Director of the Institute of Applied Physiology and Medicine, Seattle, Washington, U.S.A.:
"Every few years a dissertation comes to the area of clinical application of medical technology which carries us forward as on a magic carpet into new regions of understanding and patient care. This book is such a magic carpet. It brings together, in a clear and incisive fashion, important hemodynamic principles with a simple non-invasive method of application to a part of the cerebral vasculature which has been relatively inaccessible. To the lucky and perceptive person who reads this book, a feeling of excitement and hope for progress is engendered. The diligent application of the potentials of transcranial Doppler ultrasound brings new power to our efforts in understanding the cerebral circulation and the causes, treatment and prevention of cerebrovascular disorders."

Prices are subject to change without notice

Springer-Verlag Wien New York

Moelkerbastei 5, A-1011 Wien; Heidelberger Platz 3, D-1000 Berlin 33;
175 Fifth Avenue, New York, NY 10010, USA; 37-3, Hongo 3-chome, Bunkyo-ku, Tokyo 113, Japan